Freedom 7
The Historic Flight of Alan B. Shepard, Jr.

Other Springer-Praxis books of related interest by Colin Burgess

NASA's Scientist-Astronauts
with David J. Shayler
2006
ISBN 978-0-387-21897-7

Animals in Space: From Research Rockets to the Space Shuttle
with Chris Dubbs
2007
ISBN 978-0-387-36053-9

The First Soviet Cosmonaut Team: Their Lives, Legacies and Historical Impact
with Rex Hall, M.B.E.
2009
ISBN 978-0-387-84823-5

Selecting the Mercury Seven: The Search for America's First Astronauts
2011
ISBN 978-1-4419-8404-3

Moon Bound: Choosing and Preparing NASA's Lunar Astronauts
2013
ISBN 978-1-4614-3854-0

Colin Burgess

Freedom 7

The Historic Flight of Alan B. Shepard, Jr.

Colin Burgess
Bonnet Bay
New South Wales, Australia

SPRINGER-PRAXIS BOOKS IN SPACE EXPLORATION

ISBN 978-3-319-01155-4   ISBN 978-3-319-01156-1 (eBook)
DOI 10.1007/978-3-319-01156-1
Springer Cham Heidelberg New York Dordrecht London

Library of Congress Control Number: 2013944297

© Springer International Publishing Switzerland 2014
This work is subject to copyright. All rights are reserved by the Publisher, whether the whole or part of the material is concerned, specifically the rights of translation, reprinting, reuse of illustrations, recitation, broadcasting, reproduction on microfilms or in any other physical way, and transmission or information storage and retrieval, electronic adaptation, computer software, or by similar or dissimilar methodology now known or hereafter developed. Exempted from this legal reservation are brief excerpts in connection with reviews or scholarly analysis or material supplied specifically for the purpose of being entered and executed on a computer system, for exclusive use by the purchaser of the work. Duplication of this publication or parts thereof is permitted only under the provisions of the Copyright Law of the Publisher's location, in its current version, and permission for use must always be obtained from Springer. Permissions for use may be obtained through RightsLink at the Copyright Clearance Center. Violations are liable to prosecution under the respective Copyright Law.
The use of general descriptive names, registered names, trademarks, service marks, etc. in this publication does not imply, even in the absence of a specific statement, that such names are exempt from the relevant protective laws and regulations and therefore free for general use.
While the advice and information in this book are believed to be true and accurate at the date of publication, neither the authors nor the editors nor the publisher can accept any legal responsibility for any errors or omissions that may be made. The publisher makes no warranty, express or implied, with respect to the material contained herein.

Cover design: Jim Wilkie
Project copy editor: David M. Harland

Printed on acid-free paper

Springer is part of Springer Science+Business Media (www.springer.com)

Contents

Dedication	ix
Foreword	xi
Author's preface	xv
Acknowledgements	xvii
Illustrations	xix
Prologue	xxv

1 History and development of the Mercury-Redstone program 1
    A rocket for the Cold War 1
    The first Redstones 3
    Flight testing 5
    Handing over to NASA 5
    MR-1 launch failure 12
    MR-1A flies 20
    Tribute to the Redstone 27

2 The Mercury flight of chimpanzee Ham 29
    Out of Africa 31
    Training for space 35
    Prelude to flight 41
    Good to go 41
    A troubled flight into space 46
    Spacecraft recovery 48
    Unwanted fame 53
    Final checkout of the Redstone booster 58
    A successful test flight 60
    Russia responds 63

Contents

3 NASA's first space pilot ... 65
Cheering the Pride of Derry ... 67
A family's history ... 69
Early influences ... 69
The urge to fly ... 70
Selecting the Seven ... 74
Training for space ... 77
Settling in ... 87
"My name, José Jiménez" ... 90
Decision day ... 93
Stepping up the training ... 94
Caution wins, America loses ... 96

4 Countdown to launch ... 103
Thunder over the Cape ... 103
Diminishing chances ... 106
Postponement ... 110
A day for history ... 112
Delay after delay ... 130

5 Fifteen minutes that stopped a nation ... 139
Liftoff! ... 139
First to fly ... 144
The view from space ... 148
End of weightlessness ... 152
The final hurdle ... 154

6 Splashdown! ... 159
Civilians on board ... 160
Eyewitnesses to history ... 163
From the land to the sea ... 165
The man with the camera ... 167
Calling *Freedom 7* ... 169
A hero returns ... 170
Right on the spot ... 173
Welcome aboard, Commander ... 178
Securing the spacecraft ... 186
Crewmember memories ... 189
A phone call from the President ... 191
Moving right along ... 193
Shepard's second journey ... 198
The skipper and a tight squeeze ... 199

7 A nation celebrates ... 207
Grand Bahama Island ... 207
Results of post-flight medical examinations ... 211
Reactions abroad ... 213

	To Washington	214
	A Capitol press conference	218
8	**Epilogue**	225
	The spacecraft	225
	The astronaut	230
	An authentic American hero	233
	Fly me to the Moon	238
	A space flight legend remembered	240

Appendices 245

About the author 259

Index 261

*This book is fondly dedicated to the memory of a remarkable man from "slower, lower Delaware" (as he called his beloved state) who was respectfully acknowledged in my two previous books as a great friend, a tireless helper, a proud patriot, and a patient mentor to me; a kind-hearted man known simply to one and all as 'Sully'. His wisdom, guidance and friendship will be forever cherished, and sadly missed by me.
Vale, and profound thanks for a life replete with pride and nobility to:
Lt. Col. Walter B. ('Sully') Sullivan, Jr., U.S. Air Force (Ret'd.)
(1938-2012)*

Taken only weeks before his passing on 2 December 2012 after a prolonged illness, 'Sully' Sullivan (left) with the author.

Foreword

Duane E. Graveline, M.D., M.P.H., is a former U.S. Air Force flight surgeon, aerospace medical research scientist, and analyst of Soviet bioastronautics. He was selected as one of six scientist-astronauts by the National Aeronautics and Space Administration (NASA) in 1965. These days, he is a prolific writer on medical issues and science fiction subjects.

In the dark jungles of Cameroon the female chimpanzee hardly appeared to feel the sting of the dart – it appeared no more than the sting of a hornet. But within seconds her vision dimmed, her muscles became strangely unresponsive, and she plummeted to the ground. A native dragged her unconscious body to the center of a large net spread across the jungle floor. Soon her two offspring slowly made their way to her body and the trap was sprung. The two young chimps then began their long trip to Holloman Air Force Base in New Mexico where they joined a group of chimp trainees. The year was 1959. Their training for space was soon to start. One of them, designated No. 65 during the training program, would be called "Ham" after the Holloman Aviation Medical Center upon successful completion of his space flight.

On 31 January 1961, Ham's welcoming handshake after his 16 minute 39 second space flight became known to the world. Three months later (but unfortunately three weeks after the Soviets launched Yuri Gagarin), Alan Shepard was to make his historic space flight.

We in America, in seeking some means to erase the shame of being second in manned space flight, would say that Gagarin had no option for manual control, whereas Shepard was allowed some control of his vehicle, thereby giving us some justification for the claim to have been the first to demonstrate normal extremity function during weightlessness. But even here, there is some room for debate. Ham had full use of his extremities in his responses to blinking test lights during the MR-2 mission, demonstrating that use of extremities would be normal during zero gravity. So few doubts remained.

I had selected zero gravity deconditioning as my primary area of research, and I recall with amusement the dire predictions of other scientists who made headlines back then with their warnings of physiological malfunctions that would result from even short-term zero gravity exposure. To me the critical factor was time, and you could go to the literature on bed rest to get the lions' share of it – muscle weakness, bone demineralization and orthostatic intolerance. I even had a personal introduction to the deconditioning effects at the age of 10 years while sliding out of bed following nine days of bed rest for an appendectomy. After reassuring the nurse that I was fine, I would've slid to the floor had it not been for her support. Imagine, at age 10 years I had an introduction to the effects of zero gravity, my future research subject. Eighty per cent of astronauts returning from the space station would show a similar response to standing upright on Earth the first time.

But I was not concerned with deconditioning as a result of such short exposures to weightlessness as Ham's ballistic flight, or even that of Alan Shepard. Zero gravity deconditioning as a medical concern *would* come; but it would come much later.

You need to remember that in the Cold War climate of those days, one could be criticized for saying *anything* good about Soviet accomplishments. One of my senior officers was critical to the point of being caustic about my reports of Soviet progress in bioastronautics. I was assigned the role of intelligence analyst during this period. Calling me anti-American was one of the milder comments I would attract in those days, simply by reporting the truth. Prior to the Gemini 3 mission, by which time we had accrued a total of 34 orbits of manned space flight, the last thing that our team wanted to be told was that the Soviets had already achieved 292 manned orbits, and that their bioinstrumentation was surprisingly sophisticated. During this entire period we were gleaning what we could from Soviet data. Analyzing that data was my job. I will summarize the mission of Yuri Gagarin next, owing to its obvious implications for what soon followed in the United States.

On 12 April 1961, Yuri Gagarin made the world's first manned orbital flight. Its total duration from launch to landing was 108 minutes. His bioinstrumentation was the same as that of all the other cosmonauts who followed in the Vostok program: a respiratory monitor, two leads of electrocardiograms, blood pressure, precordial vibrocardiogram and galvanic skin response (GSR). The orbital plane was inclined at 64 degrees to the equator, and the initial altitudes were selected to guarantee natural orbital decay within the lifetime of the available consumables. The cabin atmosphere was of a composition and pressure equivalent to that at sea level.

The U.S. accessed Soviet biodata in real time, giving our space scientists relevant biodata throughout the mission. Lacking a frame of reference, we had no means of utilizing the precordial vibrocardiogram information or that of the GSR. But we did have electrocardiographic data throughout the flight, and this banished doubts about whether the human body could adjust to the new environment of zero gravity. In the jargon of the space age, Gagarin's heart rate and rhythm were nominal (expected) all the way. He displayed a normal sinus rhythm throughout (the electrical activity of each heart beat originated from the usual spot near the atrial sinus), with a relative tachycardia (faster) in the launch and pre-deorbit phases of the mission. The mission plan was to descend by parachute, so useful biotelemetry terminated at retrofire. We physiologists and doctors needed to hold our breath no longer. Our amazing bodies were able to adapt to zero gravity.

And the flight of *Freedom 7* on 5 May proved to be no different in its effect on Alan Shepard. His electrocardiogram was to show normal sinus rhythm all the way with nominal rates. The non-medical reader might wonder about my use of the term "normal sinus rhythm," and this is because the origin of our heart beat can vary considerably. The usual origin of our pacemaker is the wall of the right atrium. From there the electrical activity spreads across the atria to the nodal tissue at the junction of the atria and the ventricles. The pacemaker of the heart can be normal sinus, atrial, or nodal, or indeed any spot in between. Needless to say, had Gagarin or Shepard's pacemaker shifted to any spot in the heart other than the sinus, physiologists would have been concerned. It did not, so everyone was happy. On the basis of Gagarin's data we had no concerns about Shepard's ability to adapt to zero gravity, and he took his five minutes of weightlessness in his stride.

Alan Shepard: On 11 November 1923 in the mountains of Derry, New Hampshire was born a man who was to pee in his pants to an audience of spacecraft designers and launch personnel, and later hit golf balls on the Moon. A naval aviator of almost unsurpassed talent and cool daring prior to his selection as a Mercury astronaut, he had more flight hours than anyone else. In a community of the bold and bright, he stood out like a beacon. It seemed to me that on those gravel roads so common to space launch facilities, every bend in the road was a challenge to throw gravel with his Corvette. One time, NASA tracking brought us together at Vandenberg Air Force Base in California. He may have trusted me to read the medical console, but he never trusted me to take the wheel of his prize automobile. It was generally a pleasure to be with him except for the telephone calls. It seemed as if the whole world wanted to meet him and shake his hand. Since I was the one who sat with him at the restaurant, I was the one they called. I asked him how to handle them. He said they just want to talk, and I learned what it meant to have been in space – to be an astronaut.

Alan Shepard had the grin of a rascal and when, in 1961, a few months after his flight, I showed him a small photo which just begged to be sketched in charcoal, without hesitation he wrote across the bottom of the blank sketch paper, "That's the cleanest joke I know." I spent months working on that charcoal sketch.

Sometime in the 1990s, having had that sketch hanging in my home in northern Vermont for a couple of decades with only my guests to see it, I finally decided to drive down to Derry and turn it in. Having spent years absorbing all that was known of Gagarin, I was surprised at the twists and turns involved in trying to find where space memorabilia relating to Alan Shepard might be stored.

Most people on the main street of Derry just looked at me questioningly. Finally, one told me that he knew of some space papers stored in a room over the firehouse. Needless to say, I was astonished. There was no marker, no discernible memories – nothing to tell the world that this was the birthplace of Alan Shepard. Having just completed my ten years with the U.S. Air Force and my special assignment as an analyst of Soviet bioastronautics, the cosmonauts and astronauts were like a family to me. There could hardly be a child in the Soviet Union who didn't worship Yuri Gagarin. His name was everywhere and is still revered. Yet here was his American counterpart in some shelves over the

Duane ('Doc') Graveline with the pre-autographed sketch he drew of his astronaut colleague Alan Shepard. (Photo courtesy of Duane Graveline)

firehouse with no markers visible to the public three decades after his historic flight. The Soviets named their entire space complex after Yuri Gagarin, and to me these New Hampshire folks appeared to have almost forgotten their one-time favorite son. They were waiting to build a suitable structure, I was told. But thirty years? I would like to think that my visit, with my sketch and a few e-mails in hand, played a role in helping them finally to start building a suitable structure.

Now a well-marked sign off Interstate 93 directs traffic to Derry, the home town of Alan Shepard. Regardless of how large is the sign or the museum, they will be insufficient to encompass the memories of the man I remember.

Duane E. Graveline, M.D., M.P.H.,
Merritt Island, Florida, 2013

Author's preface

I once had the privilege – the very memorable privilege – of meeting Rear Admiral Alan Shepard. Sadly enough, it would be the only occasion. In 1993, under gloomy skies, an air show was held at Avalon airport outside of Melbourne, Australia, and I was there in uniform in my capacity as a Customer Service Manager with Qantas Airways to usher attendees through our 747 and 767 aircraft. I knew that special show guest Alan Shepard was to do a signing session outside of the Qantas VIP tent at a certain time, so I carefully orchestrated my break to be there 15 minutes ahead of that time.

As I'd assumed, Shepard was by himself in the private rear part of the VIP tent, sleeves rolled up and enjoying a quiet beer. I introduced myself, saying as we shook hands, "It's a great pleasure to meet you, Rear Admiral. I've been waiting quite a while to meet you." With that he looked at his watch and almost apologetically said, "Oh, how long have you been waiting?" At which I replied, "Since the fifth of May 1961." He laughed out loud. I then enjoyed a couple of precious minutes chatting with the man before he was called to face the public and sign a whole bunch of prints – curiously of the Space Shuttle undergoing flight tests mounted atop a 747. I really felt that something far more appropriate could have been found, but as he signed one for me it's a great souvenir of a wonderful day and an extraordinary person.

After he'd rolled down and buttoned his sleeves once again and walked out to the waiting line of autograph 'customers', I noticed Louise Shepard sitting quietly in a far corner of the tent, so for a few minutes we had a friendly, animated conversation about the places that she would dearly love to see in Australia.

The memories of that day came flooding back as I began work on this book, and I'll always be grateful that the opportunity to meet Alan Shepard came my way. It made the writing of his flight story so much more personal.

The adulation that swept most of the world – and particularly the United States – in the wake of his suborbital flight was something quite new and largely unexpected, with the sheer scale of it taking many by surprise. Following his post-flight reception and being presented with a NASA medal by President Kennedy at the White House, the Shepards traveled as planned to the Capitol building in an open limousine along with Vice President Johnson. The other Mercury astronauts trailed behind in other vehicles. Amazingly, it had been decided by NASA officials in Washington, D.C. not to organize any sort of showy parade for the nation's first astronaut. However, nobody had told the people of the nation's capital, who turned out in their thousands to line the streets and cheer Alan Shepard and his colleagues as they drove by in a fleet of limousines. Several thousand more had gathered at the steps of the Capitol to catch a glimpse of America's first astronaut, and he was obviously overwhelmed by the excitement and sheer patriotism displayed by the citizens of Washington, whom he acknowledged prior to eventually heading in to address a news conference. There was a further surprise in store when he made his way to the waiting microphones. All the news reporters and photographers stood and applauded as he fronted the gathered media – something almost without precedent.

By the time John Glenn orbited the Earth the following year, everyone knew what to expect post-flight, and true to predictions the nation exploded as the freckle-faced Marine enjoyed exultant parades throughout the country. He had become the latest, and one of the greatest, American heroes. The triumph of Shepard's history-making Mercury suborbital flight had to take something of a back seat to the man who had once served as his backup and who now enjoyed a celebrity status the like of which had not been seen since the days of Charles Lindbergh.

In 2011 our attention was turned once again to Alan Shepard's Mercury-Redstone flight, as we remembered the golden anniversary of sending America into space in a tiny capsule he had named *Freedom 7*. Sadly, he was no longer with us, having died of a lingering disease back in 1998.

As someone who has found fascination and enthrallment in the ongoing history of human space flight for the greater part of his life, I feel proud to be able to present this book on the flight that made Alan Shepard and *Freedom 7* famous.

Of necessity there is some biographical material on the life of Alan Shepard, but as the name of the book suggests, I've principally focused on his historic flight. For those seeking information on the life and other achievements of Alan Shepard, there is one biography that covers his entire lifetime; Neal Thompson's 2004 publication, *Light This Candle: The Life & Times of Alan Shepard, America's First Spaceman*.

Just as I feel so privileged to have met the first American to fly into space (and Apollo moonwalker), I am also grateful that I happened to be around and historically aware in an era in which we took, in Shepard's own words, "those first baby steps" into the astonishing wonderment and glory that is our universe.

Acknowledgements

It is always pleasing once a book is in manuscript form to acknowledge in print the assistance and support of all those people whose enthusiasm and kindness helped to shape the end product. This is the case now, in presenting this record of America's first human-tended flight into space. Brief though that mission was, it emphatically signaled the beginning of a grand enterprise embracing both science and exploration for the United States.

Firstly, I would like to acknowledge the bountiful help of some people who were either there as this historic mission evolved and was carried through to completion, or themselves witnessed the amazing events of 5 May 1961. Many thanks, therefore, for their information, photographs and memories to Dean Conger, Philip Kempland, Ed Killian, Wayne Koons, Larry Kreitzberg, H.H. ('Luge') Luejten, Paul Molinski, Earl Robb, Joe Schmitt, Charles Tynan, Jr., and Frank Yaquiant.

Other assistance was freely given by Susan Alexander, David and Debi Barka, Reuben Barton, Kerry Black (of the Scotsman Publications Library), Lou Chinal, Dr. Bruce Clark, Rory Cook (Science Museums Group, London), Rick DeNatale, Ken Havekotte, Ed Hengeveld, Richard Kaszeta, Tacye Phillipson (National Museums Scotland), J.L. Pickering, Eddie Pugh, Stéphane Sebile, Hart Sastrowardoyo, Norma Spencer, Julie Stanton, David Lee Tiller, and Charles Walker.

Special mention must also be made of the wonderfully supportive and ongoing help I received from Robert Pearlman and the space sleuths, experts, and enthusiasts who frequent his website, *www.collectspace.com*, on which no question ever passes unanswered and offers of assistance flow freely from people with a similar passion for all things to do with the history, present, and future of space exploration. For this Australian space enthusiast, a day never passes without checking at least once – and often more – the latest posts on this truly amazing forum.

As always, I have to thank an old friend and writing collaborator, Francis French, who readily lends an expert eye by reading through my chapter drafts on his daily train commute home from San Diego, seeking overlooked typos, grammatical errors, or missed (or misinterpreted) facts. His suggestions for adding extra information or stories are also greatly appreciated.

Thanks yet again to Clive Horwood of the Praxis team in the United Kingdom for his continuing support of my ideas for books. Similar thanks go to Maury Solomon, Editor of Physics and Astronomy, and Assistant Editor Nora Rawn, both at Springer in New York. Thanks to Jim Wilkie for his brilliant cover artwork. And of course to the man who provides that final polish to my work, the incomparable copyeditor and fellow space aficionado David M. Harland.

Thank you one and all for helping me to tell this truly amazing and inspiring story from the very beginning of the human space flight era.

Illustrations

Front cover:
Alan Shepard being winched up from *Freedom 7* to the recovery helicopter.
(Photo: NASA)

Back cover:
Backup pilots Gus Grissom and John Glenn with (right) Alan Shepard.
Liftoff for America's first manned space flight.
(All photos: NASA)

Dedication page
Author Colin Burgess with Walter ('Sully') Sullivan ix

Foreword
Duane Graveline shows his sketch of Alan Shepard xiv

Chapter 1
U.S Army personnel hoist a Redstone rocket 2
Redstone assembly line at the Chrysler plant 4
A Redstone on the launch pad 6
Wernher von Braun and the Mercury astronauts 8
Installation of a Mercury test capsule on a Redstone 9
Mating the test capsule to the booster rocket 10
Comparative illustration of Redstone boosters 11
McDonnell Aircraft Corporation's "clean room" 13
Spacecraft No. 2 14
Launch preparations for MR-1 15
Jettisoning the launch tower after MR-1 booster shutdown 17
Damaged launch tower on a nearby beach 18
McDonnell's Pad Leader Guenter Wendt 20

Schematic drawing of the Mercury-Redstone rocket	21
Spacecraft No. 2 at the Lewis Research Center	22
Launch of the MR-1A mission	24
Helicopter recovery of the unmanned MR-1A capsule	25
Recovery crew Allan Daniel and Wayne Koons	26

Chapter 2

The MR-2 capsule	30
"Subject 65," chimpanzee Ham	31
M/Sgt. Ed Dittmer with Ham	32
Chimpanzee space candidates in couches	33
Test subjects learning to move levers	34
Ham with one of his animal handlers	36
Subjects learning to push levers	36
Ham in his space container	37
The MR-2 psychomotor panel	38
Layout of the MR-2 spacecraft	39
Ham in his couch container	40
Backup chimpanzee Minnie with Ham	42
Inserting the container into Mercury spacecraft	43
Prior to hatch closure	44
MR-2 lifts off	45
Still images of Ham in flight	46
MR-2 spacecraft recovery at sea	49
Helicopter crewman George Cox snaring the capsule for retrieval	50
The recovered capsule arriving at the USS *Donner*	51
Lowering Ham's capsule onto the aircraft carrier	52
Opening the hatch	53
Ham's container is extracted from the spacecraft	54
Dr. Richard Benson examines Ham post-flight	55
Ham reaches for an apple	56
Removing sensors from Ham's body	56
Space pioneer Ham's final resting place in New Mexico	57
The author with the MR-2 capsule	57
The chimpanzee container shown in its flight position	58
McDonnell workers with the *Freedom 7* capsule	59
Launch of Little Joe LJ-1B	60
Rhesus monkey Sam	61
Launch of the MR-BD Redstone rocket	62

Chapter 3

Original knocker on the Shepards' front door	66
A winter photo of the Shepards' residence	67
David and Debi Barka's family	68
Pinkerton Academy	71

Pinkerton student group, 1938	72
Shepard in the U.S. Naval Academy	73
Scott Carpenter undergoing astronaut candidate testing	76
The Mercury astronauts at the McDonnell plant in Missouri	78
Shepard's parents and grandmother	79
Simulated weightlessness in a C-131 airplane	81
Cooper, Shepard and Glenn prepare for centrifuge training	82
The Johnsville centrifuge	83
Cooper prepares for a centrifuge run	84
Shepard with Grissom and the MASTIF trainer	85
Training with hand controls in a contour couch	86
Astronauts' nurse Dee O'Hara	88
The astronauts' quarters in Hangar S	89
Comedian Bill Dana	91
"First crew," Glenn, Shepard and Grissom	95
Redstone rocket ready for transport to the Cape	96
Booster is raised on the launch pad	97
Freedom 7 being hoisted for mating to the Redstone	98
Shepard holds a model of a Mercury capsule	99

Chapter 4

Heavy rain at the pad	104
Joe Schmitt adjusts sensors in the leg of Shepard's suit	105
Waiting to fly	107
The three pilots nominated for MR-3	108
Jack King announces a launch delay of two days	109
The disappointed news media	110
Pad 5 viewed from the blockhouse	111
Breakfast for Shepard and Glenn	113
Dr. Douglas examines Shepard's ears	113
Shepard undergoes a medical examination	114
Sensors attached to Shepard's body	115
Joe Schmitt makes final suit adjustments	116
Pre-flight suit pressurization check	117
Shepard departs Hangar S	118
Cooper briefs Shepard in the transfer van	119
Stepping down from the transfer van	120
A last look at the waiting Redstone	121
Entering the gantry elevator	122
Glenn welcomes Shepard to the White Room	123
Shepard thanks Grissom for his help	124
Assisted through the hatch into *Freedom 7*	125
Shepard maneuvers into his couch	126
Last photo of Shepard prior to hatch closure	127
The hatch is secured in place	128

Grissom peeks into the capsule's periscope 129
Dr. Douglas gives a final "OK" sign 129
The gantry rolls back 130
Chris Kraft, Walter Williams and Walter Kapryan 132
Preparing for launch in the Mercury Control Center 133
Gordon Cooper with Wernher von Braun in the blockhouse 134
An aerial view of the Pad 5 blockhouse 135
The "cherry-picker" ready for a launch pad evacuation 136

Chapter 5
The media gathers for the MR-3 launch 140
Inside the Pad 5 blockhouse 140
Moment of ignition for MR-3 142
The Redstone thunders into the sky 143
Climbing higher into the sky 145
Watching the flight unfold from the White House 146
Image of Shepard taken during his mission 148
Earth as viewed from *Freedom 7* 149
The Mercury capsule's retro-package 151
Schematic drawing of *Freedom 7*'s instrument panel 152
Photo of jettisoned drogue chute and antenna canister 155
Illustration of the flight's suborbital path to splashdown 156

Chapter 6
USS *Lake Champlain* at sea 160
Ed Killian on the USS *Lake Champlain*, February 1961 161
Briefing by the NASA Recovery Team Leader 163
The "island" of the USS *Lake Champlain* 164
Shepard with Wayne Koons and George Cox 167
Dean Conger attaches a camera to the recovery helicopter 168
Freedom 7 moments before splashdown 171
The spacecraft hits the water 171
Recovery helicopter hovers above *Freedom 7* 174
Shepard being winched up 176
Almost at the helicopter door 176
Distant view of Shepard's retrieval 177
Helicopter #44 hauls *Freedom 7* out of the water 177
Ship's crew watch the recovery 178
Lowering the spacecraft onto the carrier 179
Shepard exits the helicopter 181
The astronaut returns briefly to his spacecraft 182
Recovering his helmet 182
Shepard heads for medical checks with Dr. Strong 183
Photographer Dean Conger at work 183
Stripping off his spacesuit and bio-med sensors 184

Shepard records his first impressions of his flight	184
An excited Louise Shepard	185
Shepard's parents watch his safe recovery on TV	186
Steadying the spacecraft	187
The crew get a close look at *Freedom 7*	188
Cdr. Skidmore with the two helicopter recovery pilots	190
Shepard in a clean flight suit	191
A phone call from the President	192
The NASA team prepares to fly out	194
Shepard with NASA representative Charles Tynan	195
Shepard and Tynan peer into the spacecraft	196
Cameramen on the island structure	197
A farewell smile from Alan Shepard	198
Aircraft ready to fly over to Grand Bahama Island	199
Shepard aboard the TF-1 Trader aircraft	200
Dean Conger shakes hands with Shepard	201
Capt. Weymouth is assisted from inside *Freedom 7*	202
Freedom 7 departs the ship for Cape Canaveral	202
A recent photo of Ed and Kath Killian	203

Chapter 7

Slayton and Grissom welcome Shepard to Grand Bahama Island	208
Grissom takes Shepard to the hospital for medical checks	209
A relaxed astronaut on the day after he made history	212
Shepard's parents on the USS *Lake Champlain*	215
The Shepards meet the First Family	216
The President and the astronaut at the White House	216
President Kennedy hands Shepard the NASA medal	217
Vice President Johnson and the Shepards leave the White House	218
Cheering crowds line the Washington streets	219
A massive crowd at the Capitol	220
Shepard gives a news conference at the Capitol	221
Wearing his NASA medal with pride	222
Next to fly: astronaut Gus Grissom	223

Chapter 8

Freedom 7 on display at the Paris Air Show, 1961	226
Alan Shepard with his spacecraft	228
Freedom 7 is exhibited at the Smithsonian	229
Alan Shepard and John Glenn with Gherman Titov	230
Freedom 7 is delivered to the Science Museum in London	231
Onlookers watch as the spacecraft enters the Science Museum	232
The *Freedom 7* display in the Science Museum	233
On to the next exhibition stop in Edinburgh, Scotland	234
Freedom 7 at the U.S. Naval Academy, Annapolis	235

On loan to the JFK Presidential Library and Museum, Boston 235
Shepard in training for MR-3.. 236
The unflown *Freedom 7 II* spacecraft 237
The unofficial logo Shepard had painted on *Freedom 7 II* 238
The crew of Apollo 14 239
Alan Shepard standing on the Moon... 240
August 1971: Alan Shepard gains the rank of rear admiral... 241
Alan B. Shepard, Jr. 242
Memorial stone for Alan and Louise Shepard 242

Prologue

On 25 September 1961, President John F. Kennedy gave an impassioned address to the United Nations General Assembly in New York City, during which he presented proposals for a new disarmament program as well as warning of "the smoldering coals of war in Southeast Asia." He also called for peaceful cooperation in the new frontier of outer space.

"The cold reaches of the universe," Kennedy implored, "must not become the new arena of an even colder war." Earlier that year, in both his Inaugural and first State of the Union addresses, he had called for East-West cooperation "to invoke the wonders of science instead of its terrors. Together let us explore the stars."

With a smug arrogance, the hierarchy of the Soviet Union dismissed Kennedy's suggestion of cooperation in space exploration. They had very little incentive to join forces or feed information to an American space program that was then deemed to be lagging well behind theirs. Back then, they possessed an array of powerful boosters – designed to deliver massive nuclear weapons – which could insert huge payloads into orbit. Four years earlier, in October 1957, they had launched the first artificial satellite, followed weeks later by the first living creature to be sent into orbit – a dog named Laika. These would not be the only major "firsts" the Soviet Union achieved in what became universally known as the "Space Race" – a mammoth and incredibly expensive undertaking of resources and technological advances in order to gain the ascendancy in space exploration.

The previous Eisenhower administration, in spite of the best efforts of Majority Leader Lyndon B. Johnson to incite some measure of positive response, and despite the establishment of NASA as the nation's civilian space agency, had been accused of excessive tardiness in getting a viable American space program up and running. Additionally, that administration was accused of treating Soviet space efforts with skepticism and almost disdain. Even Republican officials had admonished Dwight Eisenhower over the Soviet Union's seemingly superior space program and what it might mean.

During the 1960 presidential campaign, Senator John Kennedy had come down hard on what he perceived to be the mounting "gap" in space technology. To him, Eisenhower's inaction symbolized the nation's lack of initiative, ingenuity, and vitality under Republican rule. Furthermore, he was convinced that Americans did not yet grasp the world-wide political and psychological impact of the Space Race, and that the dramatic Soviet efforts were helping to build a dangerous impression of unchallenged global leadership generally, and scientific pre-eminence particularly.

Kennedy narrowly won the election, and during the transition he appointed a task force under Science Advisor Jerome B. Wiesner to advise him on the national space program and recommend policies for the future. On 10 January 1961 the Wiesner Committee submitted its preliminary report, advising that without immediate action the United States could not possibly win a race to place the first human into space, even though the nation's first astronauts had already been selected and were deep into training for the first missions.

Wiesner was himself a strong advocate for utilizing unmanned probes rather than risking human lives in exploring space, but he also realized the imperative for setting immediate goals in space and achieving those goals. The committee's report stated that the United States was seriously lagging behind the Soviet Union in missile and space technology, attributing this to duplication of effort and a lack of coordination among NASA, the Department of Defense and the three military services, with each of those services competing to create its own independent space programs.

Before his first hundred days in the White House were over, President Kennedy's concern was dramatically proven correct. On 12 April 1961, a 27-year-old Soviet Air Force lieutenant named Yuri Gagarin was launched into a single orbit of the planet, becoming history's first human space explorer. This largely unexpected feat had a profound impact on a nation which had been looking forward with confidence to the imminent first flight of an American astronaut, albeit only a 15-minute ballistic or suborbital mission.

America's man-in-space program, which came to be called Project Mercury, had its origins during the middle years of the 1950s as a basic research initiative of the National Advisory Committee for Aeronautics (NACA). By the late summer of 1958 the momentum within NACA for a manned space program had increased to the point where it became a strong and viable discussion topic before various committees in Congress while the National Aeronautics and Space Act of 1958 was under serious consideration.

Prior to the passing of the Space Act on 29 July 1958, it had become evident that NACA would undergo a radical evolutionary change by becoming the nucleus of a proposed civilian space agency to be known as the National Aeronautics and Space Administration (NASA) which would be assigned responsibility for carrying out the nation's manned space flight program.

When NASA officially came into existence on 1 October 1958, the agency's first administrator Dr. T. Keith Glennan approved the setting up of Project Mercury and authorized the establishment of the Space Task Group (STG) to implement and oversee the project. Created on 5 November 1958, the STG was based at the Langley Research Center in Hampton, Virginia. As its director, Dr. Robert R. Gilruth, would later state:

"The methods by which Project Mercury was planned to be implemented were to use the simplest and most reliable approaches known and to depend, to the greatest extent practicable, on existing technology. To this end, existing ballistic missiles (the Atlas and

Redstone) were selected as the primary propulsion systems; it was planned to use a drag reentry vehicle with the entry initiated by retrorockets, with the final descent to be made with parachutes, and to plan on a water landing. As the Atlas and Redstone weren't designed originally for manned flight operation, it was necessary to provide automatic escape systems which would sense impending launch-vehicle malfunctions and separate the spacecraft from the launch vehicle in the event of such malfunction.

"Man had never before flown in space and thus it was felt desirable to include animal flights in the program to provide early biomedical data and to prove out, realistically, the operation of the life-support systems. It was considered wise to monitor the performance of the spacecraft, its systems, and its occupant, whether animal or man, almost continuously. To this end, a world-wide network of tracking, telemetry, and communications stations has been established.

"Since a new era of flight was being approached, it was planned to use a build-up type of flight-test program, in which each component or system would be flown to successively more severe conditions in order first to prove the concept, then to qualify the actual design, and finally to prove, through repeated use, the reliability of the system."

In the wake of the shock announcement that Yuri Gagarin had completed a single orbit barely weeks before an American astronaut was due to fly a suborbital mission, the realization that the Cold War enemy had beaten them onto the "high ground" of space came as a disturbing development to the American people. This was a country that had endured in the previous three-and-a-half weeks not only the Soviet space triumph, but also the fiasco of the Bay of Pigs invasion of Cuba. Deeply troubled, many Americans believed that the Soviet Union had demonstrated a so-far unrivaled ascendancy in breaching and exploring the new domain of space.

On 5 May 1961, at the start of a decade that began the practice of making people famous for fifteen minutes, a renewed sense of confidence arose and national pride was restored across America when a 37-year-old U.S. Navy commander was hurled into space atop a Redstone rocket. To many, this achievement fell somewhat short of a solid response to Gagarin's circuit of the globe, but that flight, in a tiny spacecraft named *Freedom 7*, ushered in America's participation in the gathering thrust of the Space Race.

As a scientific and historical fact, the first venture of an American astronaut into space deservedly stands on its own – it requires no embellishment. In the context of disastrous events troubling the United States, however, it had a special importance for everyday citizens with an urgent need of success. The embarrassing misadventure in Cuba, the apparent loss of Laos, and the shattering announcement by the Soviet Union of its own space achievement were wounds that hurt. The flight of *Freedom 7* gave America that "can do" sense of success once again, and reinvigorated a much-needed groundswell of national pride.

History will record that while the Soviet Union continued for a time to outdo the United States' efforts in human space exploration, NASA's achievements in human space flight and technology would soon outstrip those of the Soviet space chiefs as America pursued a new national goal set by President Kennedy, of landing a man on the Moon by the end of the decade and returning him safely to the Earth.

One of those NASA astronauts who proudly walked on the Moon during Project Apollo was also the man who set America on its audacious path towards that goal. He was U.S. Navy Commander Alan Bartlett Shepard, Jr.

1

History and development of the Mercury-Redstone program

They called the sleek, tubular rocket "Old Reliable," due to its dependability and an unsurpassed record of successfully completed launch and flight operations. Through these qualities, as history records, the Redstone rocket became the perfect choice for launching the first American into space.

Prior to being used as the booster vehicle for the early Project Mercury missions, the Redstone had undergone several years of development and testing as a medium-range, tactical surface-to-surface ballistic missile for the U.S. Army Ballistic Missile Agency (ABMA) located at the Redstone Arsenal in Huntsville, northern Alabama. Over time, the rocket had proved itself, as the nickname suggests, to be one of the most reliable large rockets ever produced in the United States.

A ROCKET FOR THE COLD WAR

A modified and enhanced descendant of Nazi Germany's deadly A-4/V-2 rocket, the Army's Redstone missile was developed through the efforts of around 120 captured ex-Peenemünde rocket engineers who, along with their families, had been transferred to the Huntsville facility after undertaking related ordnance work at White Sands, New Mexico in 1950. The move to Huntsville was met with much enthusiasm, as the isolation and desert sparseness of White Sands was in stark contrast to the greenery they had known in Germany. Once settled at the recently formed Ordnance Guided Missile Center (OGMC), they would continue their design and development research under the erudite leadership of recently appointed technical director, Dr. Wernher von Braun. They would also be joined in their work by hundreds of other research personnel from White Sands, including contract employees with the General Electric Company and a number of Army draftees possessing degrees in math, science, and engineering.

The Redstone (tracing its name, like the Huntsville arsenal, to the red rocks and clay soil abundant in that region) had begun life as one of three tactical missiles of differing size and capabilities that were undergoing rapid development by the Army in

2 History and development of the Mercury-Redstone program

U.S. Army personnel hoist a Redstone missile upright prior to a test firing exercise. (Photo: U.S. Army)

order to deliver nuclear warheads. These missiles were designated the Corporal, Hermes A3, and Hermes C1.

In October 1950, Kaufman T. Keller, then president of the Detroit-based Chrysler Corporation, had been appointed by Secretary of Defense George C. Marshall to the part-time post of Director of Guided Missiles, with a full-time officer of the armed forces as his deputy. It was a new post within Marshall's office, and he said at the time that Keller's task was to provide him with "competent advice in order to permit me to direct and co-ordinate activities connected with research, development and production of guided missiles." The creation of this office had been recommended to Marshall by the Secretaries of the Army, Navy, and Air Force. The Department of Defense said this step also had the approval of both the Joint Chiefs of Staff and the Research and Development Board [1]. In this capacity, Keller agreed to a request from the Office of the Chief of Ordnance to accelerate the Hermes C1 program, handing primary responsibility for the tactical nuclear guided missile program to the OGMC of the Redstone Arsenal on 10 July 1951. The following year, on 8 April 1952, the Chief of Ordnance renamed it the Redstone missile.

THE FIRST REDSTONES

Using their V-2 experience, and under orders from the Pentagon to develop a large tactical rocket capable of delivering a nuclear warhead a distance in excess of 200 miles, von Braun and his team manufactured and tested in-house a number of 69-foot prototype rockets before the task was handed over to a production contractor. These missiles were powered by a liquid-propellant engine developed by the Rocketdyne Division of North American Aviation that delivered a thrust of some 78,000 pounds.

Not surprisingly, given the involvement of so many German rocket engineers and technicians, the missile evolved with a number of similarities to the V-2 rocket, but with major improvements. "When completed, the Redstone represented an important advance over the V-2," wrote Von Hardesty and Gene Eisman in *Epic Rivalry: The Inside Story of the Soviet and American Space Race*. "The Redstone's warhead and guidance system, for example, was contained in a reentry vehicle that separated from the main body of the rocket (unlike the V-2, where the entire rocket body returned to Earth in one piece). The guidance system used a computer and an inertial navigation system contained in the warhead and relied totally upon onboard instruments. To reduce the missile's weight, the fuel tanks were formed by the outer surface of the rocket rather than being housed separately inside it." [2]

Following static and ground testing at Huntsville, the launch-ready missiles were to be transported to Cape Canaveral for firing from the test range. Retired engineer Allen Williams worked on the Redstone missile project. In 1952, while employed as a professor in mechanical engineering at Louisiana Polytechnic Institute in Ruston, Louisiana, he was approached by the Thiokol Chemical Corporation with an offer to take over the Redstone project, which was then experiencing some difficulties with the rocket engine. He decided to take up the challenge and relocated to the Redstone Arsenal to work with the German rocket scientists.

"When I took the job over, it was in serious trouble; everyone told me it would not be successful," Williams said. "But we managed to overcome the problems and scheduled flight tests at Cape Canaveral with four test shots of non-guided rockets. Things [in those days] were so unsophisticated. Cape Canaveral was a town of just about 500 people. We drove our test rockets through the town hidden under covers. The project was supposed to be secret, but we had to remove the stop lights in the town to let the rockets pass through. We did this with just about everyone in town standing around watching!"

As Williams recalled, facilities at the Cape in 1952 included only four concrete pads about 20 yards square and a blockhouse for observers, while launch warnings and tracking were fairly rudimentary affairs, as he cited in one example. "We sent planes out over the ocean to warn people to get away from the area. We didn't know where the rocket would go, but we thought it might have a range of 75 nautical miles [about 83 statute miles]. We sent out trace planes to follow the first shot. The rocket had dye-markers on it to indicate its impact area in the ocean. The plane couldn't find it at first, but 50 miles out in the ocean [the pilot] picked up the dye-markers.

"Three weeks later we held another launch, but the wind was much stronger than usual. We held the launch for the wind to die down. When we did launch the rocket, it took off and was carried like an arrow by the wind, parallel to the south coast of Florida; we tried to blow it up, but the destruct mechanism failed." Williams went on to become Thiokol's director of engineering in Elkton, Maryland [3].

The Redstone production assembly line at the Chrysler Corporation. The rockets pictured are the Jupiter variant of the Redstone. (Photo: Chrysler Corporation)

FLIGHT TESTING

Redstone design work was completed in 1952. In October, after the first models had been manufactured at Huntsville, the Chrysler Corporation was hired to build them in Detroit, Michigan. The contract was sealed on 19 June 1953, just five weeks prior to the armistice of the Korean War. The production home of the Redstone was to be a vast government-owned plant located in what was better recognized back then as the world's automotive center. In fact, the agreement called for the prime contractor to build the first 12 missiles at the Redstone Arsenal. The remainder were all built by Chrysler. While the total number of Redstone missiles built varies by source, there were at least 137 and perhaps as many as 146.

The first flight test of a Redstone was at Cape Canaveral on 20 August 1953, but a fault in the inertial guidance system caused it to go awry. After it had struggled to an altitude of 24,000 feet the range safety officer detonated a package of dynamite built into the wayward rocket, blowing it to pieces before it could fall back and cause any damage on the ground. With the problem identified through radio telemetry, the fault was fixed and the second flight was successfully completed.

Test flights continued over the next five years, and many refinements were made to enhance the rocket's already enviable reliability. From 1953 through 1958, a total of 37 were fired to test structure, engine performance, guidance and control, tracking and telemetry.

In August of 1958, a Redstone became the first American missile to participate in a nuclear test, by detonating a 3.8 megaton warhead as part of Operation Hardtack. While the Redstone's role as a weapon delivery system was brief, it nevertheless had a major impact on America's early space program.

HANDING OVER TO NASA

The Redstone's association with Project Mercury began as the result of a January 1958 meeting of top military personnel at the ABMA. The Department of the Army had proposed a joint Army, Navy, and Air Force program to send a man into space and back, under the project working title of "Man Very High." In April, however, the Air Force decided that it did not wish to participate, and the Navy was becoming increasingly lukewarm on the joint service venture. As a result, the Army decided to push on with the project alone. Now redesignated "Project Adam," a formal proposal for the military space venture was submitted to the Office of the Chief of Research and Development on 17 April 1958.

As outlined in the proposal, the intention of Project Adam was to send a man on a ballistic flight to an altitude of around 170 to 200 miles in a recoverable capsule atop a Redstone missile. It was pointed out that much of the supporting research for such a space venture had already been carried out at the ABMA in Huntsville. In turn, the Secretary of the Army, Wilber M. Brucker, forwarded the Project Adam proposal to the Department of Defense's recently created Advanced Research Projects Agency (ARPA) the following month, along with a recommendation that the agency consider funding the project. But this was rejected in a memorandum to the Secretary of the Army dated 11 July 1958, in which Roy W. Johnson, the first director of the ARPA, indicated that Project Adam was not considered integral to the agency's own man-in-space program.

6 History and development of the Mercury-Redstone program

A Redstone missile on the launch pad at Cape Canaveral, 16 May 1958. (Photo: NASA)

On 8 August 1958 President Eisenhower appointed 52-year-old Dr. Thomas Keith Glennan, president of the Case Institute of Technology in Cleveland, Ohio, to serve as the first administrator of NASA. For continuity his deputy would be 60-year-old Hugh L. Dryden, who had been in charge of the National Advisory Committee for Aeronautics (NACA). NASA came into official existence on 1 October. Just two weeks later, Glennan was in conflict with the Secretary of the Army about a proposal that the Army transfer to NASA the 2,100 scientists and engineers at the ABMA and all of its facilities and personnel at the Jet Propulsion Laboratory (JPL) in Pasadena, California. In the end, on 3 December, President Eisenhower ordered a compromise which involved transferring JPL to NASA, but under the direction of the California Institute of Technology. The president allowed the Army to retain its ABMA and the people under von Braun, but granted NASA the right to make use of the Huntsville capabilities on a fully cooperative basis. Glennan later announced that an increasing proportion of the work undertaken at Huntsville would be shifted to his agency in future contracts.

When NASA subsequently sought discussions on the possible use of the Army's Redstone or Jupiter rockets in support of the civilian manned space program, the Army, now without a manned space program due to the decision to abandon Project Adam, decided to cooperate. As a result, NASA issued a request to the ABMA for eight Redstone missiles to be used by Project Mercury. By arrangement with NASA, these rockets were to be assembled by the Chrysler Corporation and shipped to the Redstone Arsenal for final checkout by the ABMA.

At the time of the Redstone's selection for the Mercury program in January 1959, there were two very different versions of the rocket. The first, designated Block II, was an advanced version of the tactical missile design incorporating an improved engine, the Rocketdyne A-7, which used a combination of alcohol and liquid oxygen (LOX) as its propellants. However, there were concerns that this propellant mixture would almost – but not entirely – achieve the thrust necessary to boost a one-ton spacecraft into space. North American Aviation set its Rocketdyne Division the task of increasing the thrust by about 5 percent with the proviso that there could not be any changes to the existing propellant systems. They finally came up with a toxic mixture of unsymmetrical dimethylhydrazine and diethylenetriamine, to which the military gave the name Hydyne. When combined with LOX, this would provide the required additional thrust. The Hydyne/LOX propellant was successfully utilized by the second modified version of the Redstone. Designated the Jupiter-C, this was a multistage rocket with larger tanks and Rocketdyne's A-5 engine. It was a four-stage version of the Jupiter-C with small solid-fuel rockets for the upper stages that placed America's first satellite, *Explorer 1*, into orbit around the Earth on 31 January 1958.

On 1 July 1960, a section of the Redstone Arsenal was transferred to NASA and a few weeks later President Eisenhower and NASA Administrator Glennan attended the rededication ceremony that made it the George C. Marshall Space Flight Center, which rapidly became known by the acronym MSFC. It was here that a number of Redstone rockets would be produced and tested for the civilian space agency by the Development Operations Division of the ABMA in collaboration with the project management of the Space Task Group (STG). Cooperative panels were established between Marshall, the STG, and the McDonnell Aircraft Corporation of St. Louis,

The seven Mercury astronauts visit the Fabrication Laboratory of the Development Operations Division at the Army Ballistic Missile Agency (which was renamed the Marshall Space Flight Center) in Huntsville. From left: Gus Grissom, Wally Schirra, Alan Shepard, John Glenn, Scott Carpenter, Gordon Cooper, Deke Slayton, and Dr. Wernher von Braun. (Photo: NASA, MFSC)

Missouri, which was manufacturing the Mercury spacecraft, in order to implement standards, to coordinate design and operational goals between the three agencies, and to seamlessly integrate any changes into the overall program.

The Redstone configuration selected to meet the performance requirements of the Mercury program coupled the A-7 engine and propellants of the Block II model with the enlarged tankage of the Jupiter-C. In order to support the objectives of Project Mercury, some 800 modifications were made to the rocket's existing characteristics and performance. This included the elongation of the 70-inch diameter tank unit by six feet to hold the fuel required for an additional 20 seconds of engine burn time. A new instrument compartment and adapter section was manufactured to mate with the spacecraft, along with an abort system developed by MSFC to protect the capsule and, eventually, its human occupant. With the capsule and escape power mounted on top, the Mercury-modified Redstone now stood at an overall length of 83 feet. The total liftoff weight at launch would be 66,000 pounds.

Preparing a Mercury test capsule and escape system for a Redstone test circa 1960 at the Marshall Space Flight Center. (Photo: NASA/MSFC)

A close view of the test capsule and escape system. (Photo: NASA/MSFC)

Dr. Joachim Küttner was one of Wernher von Braun's team of German scientists, and he said at the time that a great deal of care was taken in the production of each Redstone booster. "As these thin sheets of aluminum are curved and welded, each seam is minutely inspected by X-ray to make sure there are no invisible flaws that might give way under the extreme stresses of flight. Every component we use bears a special seal representing the winged god Mercury. This symbol constantly reminds assembly-line workers that a man's life depends on the product." [4]

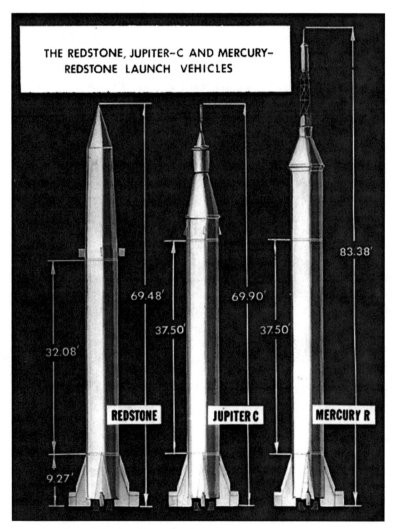

A comparative illustration of the Redstone, Jupiter-C, and Mercury-Redstone launch vehicles. (Photo: NASA, MSFC)

MR-1 LAUNCH FAILURE

Mercury-Redstone 1 (MR-1) was to be the first qualifying flight of an unmanned Mercury capsule mated with a Redstone rocket. The launch, set to take place from Pad 5 of Cape Canaveral's Air Force Launch Complex 56, was to be a full test of the spacecraft's automated flight controls, as well as the launch, tracking and recovery operations on the ground. It was also intended to provide a test of the Mercury-Redstone's automatic in-flight abort sensing system, which would be operating in an "open loop" mode. Basically, this meant that because it was an unmanned test of the system and there were no real safety issues, an abort signal would simply be ignored in order to prevent a false signal from terminating the flight needlessly.

On 22 July 1960, after the capsule systems tests in St. Louis, Missouri, Mercury Capsule No. 2 was shipped to Huntsville, where the one-ton capsule was mated with its Redstone. Both then underwent a series of checks to ensure their compatibility. Next, the total assembly, now some 83 feet long, was shipped to Cape Canaveral, arriving on 24 July. The escape tower was then prepared, and hangar activities such as the installation of parachutes and pyrotechnics were completed in time to transport the spacecraft to the launch pad on 26 September, where the rocket, enclosed by the service structure gantry, was waiting.

The first launch attempt was set for 7 November, with the intent of hurling the capsule about 220 miles over the Atlantic into a target area northwest of Grand Bahama Island. Things went smoothly until a problem caused the test to be canceled 22 minutes prior to the planned time of liftoff. According to 'Luge' Luetjen from the McDonnell Aircraft Corporation, who was then serving as the company's Redstone Mission Capsule Controller, "It was noted that the helium pressure in the spacecraft control systems had dropped below the acceptable level. A leak in the system, unfortunately under the heat shield, was obvious, and as a result the launch was scrubbed. The spacecraft was removed from the booster and the heat shield dropped to expose the culprit, a leaky relief valve. It, and a toroidal hydrogen peroxide tank were replaced, plus a minor wiring change was made as the result of an earlier test at Wallops Island, Virginia. The spacecraft was reassembled, remated to the booster, and appropriate tests rerun in order to confirm the spacecraft's integrity and that the problem had indeed been fixed." [5]

A second launch attempt was scheduled for 21 November. Two days after the 7 November launch scrub, Senator John F. Kennedy narrowly beat Richard M. Nixon, the incumbent vice president, in the election to become the 35th President of the United States. It would prove a momentous victory in terms of space flight history.

On hand to observe the second attempt to launch MR-1 – as indeed for the first – were the seven Mercury astronauts. That morning there was only a minor one-hour delay to enable technicians to fix a leak in the capsule's hydrogen peroxide system, which slipped the time of liftoff to 9.00 a.m. (EST). There were no further delays. Right on the hour the firing command was issued from the Mercury Control Center. The booster ignited, then one second later the Rocketdyne A-7 engine unexpectedly shut down. During that interval, the booster had actually lifted a little less than four inches off its pedestal. After the engine cut off, the vehicle settled back down onto its

The "clean room" at the McDonnell Aircraft plant in St. Louis, Missouri, where the Mercury capsules took shape. Extreme precautions were taken to prevent dust and metal particles infiltrating sensitive areas. In the upper photo, Spacecraft No. 2 is to the fore; it would be utilized on two Mercury-Redstone flights. (Photo: McDonnell Aircraft Corporation).

14 **History and development of the Mercury-Redstone program**

Spacecraft No. 2 is hoisted prior to being mated with the waiting Redstone rocket in Huntsville. (Photo: NASA)

On 21 November 1960 preparations continue for launching the MR-1 mission from Cape Canaveral. (Photo: NASA)

fins, slightly deforming their frames. Incredulous controllers in the blockhouse could only watch as the Redstone wobbled after the set-down impact, still venting liquid oxygen. Fortunately the 66,000-pound assembly managed to remain upright and did not explode.

Compounding the problem, the engine cutoff had initiated the emergency escape system, which activated the escape tower and recovery sequence. The escape tower's engine suddenly ignited, releasing it from the Mercury capsule and sending it soaring over 4,000 feet into the air. Eventually, it crashed down on a beach 400 yards from the pad. As stated by NASA in *This New Ocean: A History of Project Mercury*, three seconds after the escape rocket separated, "the drogue parachute shot upward, and then the main chute spurted out of the top of the capsule, followed by the reserve, and both fluttered down alongside the Redstone." [6] The radio antenna fairing was also ejected in the process.

Clearly, the only thing that would launch that day was the escape tower. One can only imagine the feelings of the astronauts as they witnessed the entire mishap.

Securing the slightly wrinkled booster and the still firmly attached capsule had to be carried out with the utmost caution, as the booster's destruct system could not be disarmed until the battery that powered it had fully depleted, which was not until the next morning. Also, the spacecraft was still on internal power and its pyrotechnics – including the posigrade and retrograde rockets – were still armed. Furthermore, it was not possible to open the vent valves and undertake the defueling process. It was therefore necessary to wait until the Redstone's liquid oxygen had fully evaporated, which would take 24 hours. The other worrying safety issue was the main parachute, which was dangling from the top of the capsule. Any strong gust of wind could cause the canopy to billow and topple the vehicle off its pedestal. Fortunately, the weather remained calm.

The following day, Walter F. Burke of McDonnell volunteered to lead a squad of men to disarm the pyrotechnics and other immediate problems [7]. The liquid oxygen tank was vented, as were the high-pressure nitrogen spheres in the pneumatic system of the engine. The fuel and hydrogen peroxide tanks were then emptied. All circuits were deactivated, the service structure was rolled back in, and finally the booster and capsule disarming was finished. The next afternoon NASA's Chief of Manned Space Flight, George M. Low, said that the MR-1 failure was believed to have been caused by the premature disconnection of a booster tail plug.

According to *The Mercury-Redstone Project* issued by the Marshall Space Flight Center in September 1961:

> The investigation which followed found the cause of the engine shutdown to be due to a "sneak" circuit created when the two electrical connectors in Fin II disconnected in the reverse order. Normally the 60-pin control connector separates before the 4-pin power connector. However, during vehicle erection and alignment on the launch pedestal, a tactical Redstone control cable was substituted for the specially shortened Mercury cable. The cable clamping block was then adjusted, but apparently not enough to fully compensate for the longer Redstone cable.

MR-1 launch failure 17

As the booster umbilical continues to fall away, the Redstone's engine suddenly cuts off, triggering the solid-rocket jettison of the launch tower. (Photo: NASA)

18   **History and development of the Mercury-Redstone program**

The escape tower plowed into beach sand some 400 yards from the pad. (Photo: NASA)

Because of the improper mechanical adjustments, the power plug disconnected 29 milliseconds prior to the control plug. This permitted part of a three-amp current, which would have normally returned to ground through the power plug, to pass through the "normal cutoff" relay and its ground diode. The cutoff terminated thrust and jettisoned the escape tower [8].

Project Mercury Director Robert R. Gilruth added that a faulty electrical circuit between the ground apparatus and the booster had simulated a normal booster cutoff signal. Ordinarily, this would be given in flight, after the booster had been taken to its designed speed and altitude of about 40 miles. The normal sequence was then for the escape-rocket tower to separate; the recovery devices such as parachutes to be armed ready for deployment; and, when the booster's thrust tailed off, the booster/capsule securing ring to be released. The separation of the escape tower took place normally, but the sea level atmospheric pressure led to the deployment of the drogue and main parachutes, and the weight of the capsule on the booster prevented its separation [9].

The late Guenter Wendt was a German-American engineer who had emigrated to the United States in 1949. He found ready employment with the McDonnell Aircraft

Corporation, and later became well known for his work in America's space program. As the man in charge of the spacecraft close-out crew on the pad, he soon gained the respect of the astronauts. To the company his title was Pad Leader, but the astronauts jokingly (and affectionately) enjoyed referring to him as their "Pad Führer," and he fell in with their humor. Wendt had been at the Cape for the failed MR-1 launch, and described the follow-on events as he saw them during a 2001 interview with Francis French.

"When we started off, we had one Redstone that lifted off about four inches and set back down. It had a little kink in it, and we could not depressurize the tank – the tank was building up pressure. I go back to the blockhouse, and the next thing I hear are [Kurt] Debus and John Yardley discussing it. Debus tells the pad safety officer to call the base and get some guns, [because] he is going to shoot holes in the oxygen tank to relieve the pressure! John Yardley says, 'Like hell you do! I have a perfect, safe spacecraft out there, it's the only one I have right now. If you shoot holes, the thing is going to blow up and I'll have no spacecraft!'

"So, what do we do? We get our engineers together to see what we could do to disarm the rocket. The first thing is, we have to get rid of the pressure in the oxygen tank. How can we do that? One of the ways is to send a mechanic out there, into the tail end of the rocket, to hook up a quarter-inch nitrogen line, then open up a hand valve. However, we don't know what will happen, so when we open it, we run like hell back to the blockhouse! A guy by the name of Sonny came, he went out, opened it, ran like hell, and just about hit the blockhouse when a big stream of gas, tens of feet long, came out. But nothing blew up.

"Next, Yardley called me and said, 'We've determined that, due to the sequencing – the main chute came out, the tower had left – the sequencer is looking at a half-G switch. When that thing activates, it will fire the retrorockets into the oxygen tank!' So now, we are looking for someone to go out there and deactivate the circuitry. However, since the periscope had retracted, you have to drill out a bunch of rivets and open the periscope door, because the electrical umbilical plug is under it. Then four jumper wires have to be plugged in. So we were looking for people with no dependents to volunteer. If the retrorocket had fired, that would be it.

"After a long discussion, some of us volunteered to go out and do it. Before we did, we had Pad Safety set up a movie camera next to the blockhouse. If it blew, at least we'd know where the pieces went! We went up there. On the Mercury capsule, the hatch was bolted down with screws – you could move the washer, but the screws had to be tight. They had to be just matched. They needed to expand on the outside, because of the heat. I had a guy who had meticulously matched each screw to a perfect hole, and stored them on foam. We got up there, got the screws out and pitched them behind us. I will never forget that. We thought, he will kill us when he finds out what happened to his matched screws! We got the hatch open, found the two switches: click, click, and we were safe. We saved the spacecraft, though we needed a new booster.

"The people who made the decisions were right there, and they made the decisions. That's what we got paid for!" [10]

McDonnell Aircraft Corporation's Pad Leader, Guenter Wendt. (Photo: NASA)

MR-1A FLIES

Within a week, a new test flight had been scheduled and designated MR-1A. While a replacement Redstone rocket would be used, it was felt that since Spacecraft No. 2 was still in good condition, after a little renovation it could be reused on the MR-1A mission with an antenna fairing borrowed from another capsule and straddled by a replacement escape tower.

Although some damage had occurred to the Redstone booster's tail assembly, engineers agreed that it could be refurbished. It was therefore crated up and shipped off to the Marshall Space Flight Center in Huntsville, where it was held in reserve until the conclusion of the Mercury-Redstone program. However, the MR-1 rocket would never actually be used and was placed on display at Space Orientation Center there.

On 8 December 1960, Spacecraft No. 2 was hoisted upward for a second mating with a Redstone launch vehicle at Cape Canaveral. It was essentially the same 2,400-pound capsule, apart from a few replacement parts and some minor modifications in areas such as the launch escape tower and the parachute deployment system.

As before, the pre-flight testing proceeded very smoothly and, with everything in order, the launch was set for 19 December. Early that morning, strong winds gusting to 150 knots aloft obliged a 40-minute hold. Next, a leak in a high-pressure nitrogen peroxide solenoid valve in the capsule caused another delay of 3 hours 15 minutes [11].

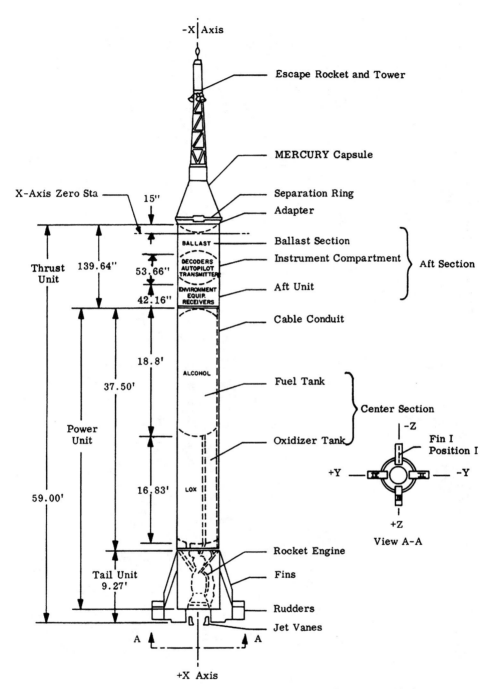

A schematic drawing of the Mercury-Redstone launch vehicle. (Photo: NASA)

Spacecraft No. 2 at the Lewis Research Center prior to its move to Cape Canaveral. (Photo: NASA)

As on previous launches, the seven Mercury astronauts somewhat apprehensively looked on as the 83-foot stack of the escape tower, spacecraft, and booster lifted off from Pad 5 of Launch Complex 56 at 11.15 a.m. (EST). Two of the astronauts, Deke Slayton and Gus Grissom, were observing from the cockpits of their airborne F-106 jets, ready to follow the ground track of the Redstone for a short time and hopefully photograph the capsule descending on its parachute over the recovery area.

This time everything went smoothly; following the ignition command issued from the blockhouse, smoke billowed from beneath the rocket and MR-1A lifted slowly off its pedestal into a clear sky, accelerating as it climbed. A brilliant trail of flame traced the sleek Redstone's course as it streaked up and tilted toward the southeast, out over the Atlantic Missile Range. Seconds later, Slayton and Grissom ripped over the Cape in their F-106s, flying in the same direction at nearly twice the speed of sound. Observers at the Cape could just make out the booster shutdown and capsule separation 143 seconds after launch.

Throughout the capsule's flight all of the systems functioned well, although the booster's velocity was 260 feet per second faster than expected at around 4,200 miles an hour, causing it to ascend seven miles higher than the predicted 128 miles. This, and high tail winds of almost 100 m.p.h., caused the separated spacecraft to travel 15 miles further downrange than expected. NASA said the bell-shaped capsule floated down by parachute into the ocean about 16 minutes after liftoff. It was first spotted approximately 90 miles northeast of Grand Bahama Island and eight miles from the prime recovery vessel, the aircraft carrier USS *Valley Forge* (CV-45).

The helicopter recovery pilots were Lt. Wayne Koons of Lyons, Kansas and Capt. Allen Daniel, Jr., of Greenwood, Mississippi. Both were members of Marine Air Group 252, which was based at Jacksonville, North Carolina. Their H-34 left the *Valley Forge* and flew over to the floating capsule, hooked on to its recovery loop, and hoisted it from the sea at 11.46 a.m., 31 minutes after it was launched. They flew back to the ship with their precious cargo and carefully deposited it on the carrier's flight deck at 12.03 p.m.

Following the successful recovery operation, the *Valley Forge* steamed to a point off Cape Canaveral within several hundred yards of the test center, then Koons and Daniel lifted the capsule and delivered it to the test center. It would later be taken to Langley Field, Virginia to be studied by technicians, engineers and scientists.

A preliminary examination revealed only minor damage to the spacecraft. The painted letters "United States" on the side had been slightly scorched by the 600-degree heat of reentry. One of the three thicknesses of glass on a small side porthole was broken, but a NASA official suggested to reporters that this could have occurred during the recovery operation. As if to demonstrate it was still functioning well after its flight into space, a bright flashing light designed to aid recovery still winked atop the nine-foot capsule.

The Director of NASA's Marshall Space Flight Center, Wernher von Braun, was delighted by the successful flight, and said that "everything was right on the money." Meanwhile Robert Gilruth, in charge of the Space Task Group, called the launch an "unqualified success." However, he cautioned that it did not indicate an immediate

A successful launch begins the MR-1A mission. (Photo: NASA)

After a successful recovery, the unmanned MR-1A capsule is safely deposited on the deck of the USS *Valley Forge*. (Photo: Associated Press)

On 19 December 1960 U.S. Marine helicopter crew Capt. Allen K. Daniel, Jr. (left) and 1st. Lt. Wayne Koons plucked the unmanned MR-1A capsule from the Atlantic after a successful 16-minute ballistic test of its systems. (Photo: Associated Press)

readiness to send a man into space. He said more flights would be needed to qualify the reliability and operation of the system, and that the next launch, expected within a month or two, might carry a chimpanzee [12].

The performance required of the Redstone rocket for the first phase of the manned space flight program had been established. It had demonstrated both the reliability and the performance needed to place the Mercury spacecraft safely into a suborbital trajectory. However, as McDonnell Pad Leader Guenter Wendt pointed out, even as their proficiency and confidence grew in safely launching rockets, there remained a great many lessons to be learned.

"All the rules changed quite a bit. At the same time, there was a lot of stuff we just plain didn't know. No one had done it before. For example, we had an escape rocket on top of the capsule. It was neatly protected with plastic that we had wrapped around it. It was great to keep the rain out. Then one day we had some Air Force people who had a satellite in a spin test facility – they spun the satellite while it was wrapped in plastic, then upwrapped it – and the satellite blew up. Static electricity. The Air Force told us the kind of plastic they'd used. It was the same kind I used on the escape rocket. Whoops! This is when you learn the hard way." [13]

TRIBUTE TO THE REDSTONE

Alex McCool began working on the Redstone engine program at Huntsville in 1954, and six years later he joined NASA in order to continue working with Dr. von Braun on the development of larger launch vehicles, including the mighty Saturn rockets. In later years he became manager of the Space Shuttle Projects Office at Marshall, and in a 2003 interview for the *Huntsville Times* to commemorate the 50th anniversary of the launch of the first Redstone rocket, McCool was asked to reflect on the early days of the rocket known as "Old Reliable."

"Really, that rocket, and the propulsion work that went into it, was the beginning of space for us here," said McCool. "There wouldn't have been a space program, or a Space Shuttle or a trip to the Moon without the Redstone. It's the foundation of what we do today. It was the beginning of the space program for America.

"The Germans had been working on other advanced rockets after they developed the V-2, and the Redstone used a lot of that work. They brought a lot of that material with them. They had been working successfully on rockets throughout the war.

"Early on, von Braun had thought about going into space," McCool reflected. "He talked about it in public all the time, and the Germans had been working on rocket designs for it. He had worked out plans to modify the Redstone to carry a man early on, in the mid-1950s, while still working for the Army. They'd talked about putting somebody up in space even then." [14]

References

1. *St. Petersburg Times* (Florida) newspaper article "Keller Named Director of Guided Missiles," issue 26 October 1950, pg 2
2. Hardesty, Von and Gene Eisman, *Epic Rivalry: The Inside Story of the Soviet and American Space Race*, National Geographic, Washington, D.C., pg. 50
3. *Leader-Post* (Saskatchewan) newspaper article, "He watched space trip with real satisfaction," issue 12 August 1975, page 12
4. Weaver, Kenneth, *National Geographic* magazine article "Project Mercury: Countdown for Space," issue May 1961 (Vol. 119, No. 5), pp. 705–6
5. Luetjen, H.H., *When Mercury Rose: The Half Life of an Ex-Spaceman*, Inter-State Publishing, Sedalia, Missouri, 2001

6. Swenson, Loyd S, Jr., James M. Grimwood and Charles C. Alexander, "MR-1: The Four-Inch Flight," from *This New Ocean: A History of Project Mercury*, NASA SP-4201, Washington, D.C., 1989
7. Ibid
8. *The Mercury-Redstone Project, Section 8, Flight Test Program*, NASA George C. Marshall Space Flight Center publication TMX 53107, Huntsville Alabama, September 1961, pp. 8–3/8–5
9. *Flight International* (U.K.) magazine article "Mercury Success," issue 30 December 1960, pg. 1006
10. Interview with Guenter Wendt conducted by Francis French, 20 October 2001, Los Angeles, California.
11. *The Mercury-Redstone Project, Section 8, Flight Test Program*, NASA George C. Marshall Space Flight Center publication TMX 53107, Huntsville, Alabama, September 1961, pg. 8–6
12. *Eugene Register* (Oregon) newspaper article, "U.S. Closer to Manned Rocket," issue 20 December 1960, pg. 2
13. Interview with Guenter Wendt conducted by Francis French, 20 October 2001, Los Angeles, California.
14. Spires, Shelby G., article "Redstone Rocket Turns Golden Today," *The Huntsville Times* newspaper, 20 August 2003

2

The Mercury flight of chimpanzee Ham

By 31 January 1961, the United States was a nation undergoing radical cultural and ethical upheaval. Changes were swirling in the wind. On that day James Meredith, an African-American, applied for admission to the all-white University of Mississippi, known as "Ole Miss," and so began a hard-fought legal action that would end in the desegregation of the university and the post-graduation shooting and wounding of Meredith by a white supremacist. That same day, a federal district court ordered the admission of two black students into Georgia University, and the State of Georgia repealed its long-standing laws which segregated the races in its public schools. The university was subsequently desegregated.

Also on that memorable date in American history, space science was on the verge of taking a huge leap forward as a Redstone rocket stood fully fueled on launch pad LC-5 at Cape Canaveral. All was in readiness for the launch of a suborbital mission designated Mercury-Redstone 2 (MR-2). It was hoped that this flight would provide the first major test of several new designs in the Mercury spacecraft, including the environmental control system (ECS), as well as a pneumatic landing bag intended to absorb much of the impact shock when a returning capsule hit the water.

But this time, as America prepared to send a man into space, there was a fully trained passenger on board the spacecraft, namely a 37¼-pound chimpanzee. NASA has always had qualms about giving personable names to animals involved in space research missions lest there be fatal accidents, so during the flight training process – as with his fellow chimps – this one was only supposed to be identified as "Subject 65." He had been allocated this number instead of the mildly offensive "Chop Chop Chang" by which he had been known early in his training, but to his handlers he was unofficially called Ham.

Immediately after his safe recovery, the chimpanzee would be publicly identified in the agency's press releases not by his subject number, but as Ham. According to popular history, this name was derived from the acronym for the Holloman Aero Medical Research Laboratory, but as his chief handler, M/Sgt. Edward C. Dittmer,

The MR-2 capsule undergoing finishing work at the McDonnell Aircraft Corporation in St. Louis, Missouri. (Photo: McDonnell Douglas Corporation)

"Subject 65," also known as Ham. (Photo: NASA)

wryly pointed out to the author, "Our lab commander at that time was a Lieutenant Colonel Hamilton Blackshear, whose friends called him Ham, so there may have been a dual purpose behind that particular name."

Dittmer also revealed that he enjoyed a great relationship with Ham. "He was wonderful: he performed so well and was a remarkably easy chimp to handle. I'd hold him and he was just like a little kid. He'd put his arm around me and he'd play ... he was a well-tempered chimp." [1]

OUT OF AFRICA

Ham was a *Pan troglodyte* chimpanzee, said (through dental analysis) to have been born around July 1957, and was one of several animals captured by trappers at a very young age in the dense tropical rainforests and savannah of the French Cameroons in Equatorial Africa. According to an article in the April 1962 issue of *The Airman*, three members of the U.S. Air Force had flown to the French Cameroons to pick up a number of animals.

The Mercury flight of chimpanzee Ham

M/Sgt. Ed Dittmer assisted Ham in his flight training. (Photo courtesy of Edward C. Dittmer)

As one of these men recalled, "When the chimps were captured, they were very small and usually ranged in age from 10 to 18 months. The natives tie them with strips of bamboo when they capture them, and make no particular arrangements for holding or feeding the young animals. When the vendor, who sells them to us, finally obtains them, they are quite heavily parasitized and malnourished." [2]

Following their transportation to the United States in 1959, Ham and the other young chimpanzees were temporarily housed at the now-defunct Rare Bird Farm in Miami, Florida. Eventually this latest batch of chimps was delivered to Holloman AFB's Air Development Center in New Mexico to join an established colony, where they were assigned identifying subject numbers and unofficial training names such as Caledonia, Chu, Duane, Elvis, George, Jim, Little Jim, Minnie, Paleface, Pattie, Roscoe, and Tiger.

Dittmer was one of several aeromedical technicians assisting in bioastronautics research for the Air Force Systems Command at Holloman AFB, reporting directly to Capt. David Simons at the Space Biology Branch of the 6571st Aero Medical Research Laboratory.

Another member of the Holloman research team was Dr. James P. Henry, who had earlier been involved in studies of blood action under heavy gravity weights and had conducted pioneering work in developing high-altitude protective clothing. Dr. Henry

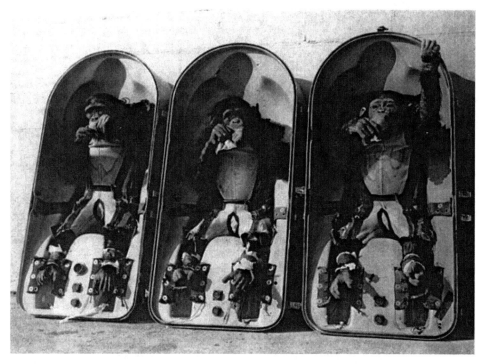

Chimpanzee space candidates Duane, Jim, and Chu enjoy a snack while training to endure prolonged periods strapped into a capsule couch at Holloman AFB. (Photo: USAF)

was appointed as an Air Force representative to a NASA committee charged with defining and setting in motion plans and procedures for animal flights within Project Mercury. He was assigned the role of coordinator for these flights under Lt. Col. Stanley White, a physician and the leader of the Mercury medical team, and he became part of NASA's Space Task Group at Langley Field, Virginia.

Henry's specific responsibilities included the establishment of an animal flight test protocol, developing the operational flight plans, and overseeing the design and manufacture of the flight hardware. He would also monitor the chimpanzee regime at Holloman, where personnel from the research laboratory had been training animals for space flight since July 1959. Initially, the plan was to train and test ten suitable chimpanzees from the colony. As with earlier programs, they began by incrementally conditioning the animals to accept the restraint conditions to which they would be subjected in a spacecraft [3].

Ed Dittmer became involved in working with the chimpanzee colony under the Space Biology Branch at Holloman, where he was the officer in charge. "Back then we got these small chimps from Africa – they were about a year old – and we started a training project," he recalled. "Of course a lot of things were classified back then, so

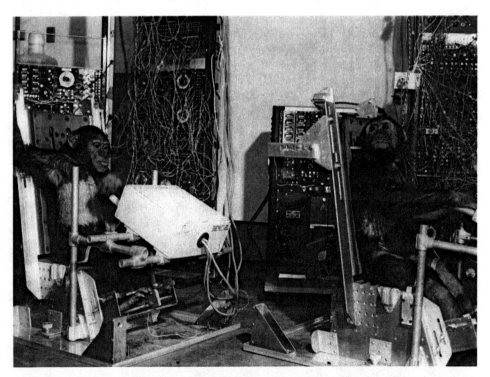

The test subjects had to learn to sit in metal chairs and move levers. Ham is seated at the rear; the chimp at front is Enos, who would fly an orbital mission in November 1961. (Photo: NASA)

we had no real idea what we were training these chimps for, but we were teaching them to sit up and work in centrifuges, so it was quite evident that we were training them for use in missiles.

"We started out by teaching them to sit in these little metal chairs, set about four or five feet apart so that they couldn't play with each other. We'd dress them in little nylon web jackets which went over their chests, and then fasten them to their chair. We'd keep them in the chairs for about five minutes or so and feed them apples and other fruit, and we'd progressively put them in their seats for longer periods each day. Eventually they'd just sit there all day and play quite happily." [4]

Each of the chimpanzees was kitted out with one of these nylon "spacesuits," and soon came to accept wearing them. During lengthy training exercises, a diaper would also be worn beneath the nylon suit.

TRAINING FOR SPACE

After the chimpanzees had become familiar with sitting in the steel chairs, Dittmer's team began securing them in individually molded aluminum couches. These were smaller versions of contour couches that the astronauts would one day occupy in the Mercury spacecraft. Next, the animals were introduced to a device mounted across their lap that was called a psychomotor, a small machine specifically designed to test their reflexes and responses.

Apart from participating in tests of the spacecraft's life support systems, one of the main tasks that the MR-2 chimpanzees had to master was pushing levers on the psychomotor in sequence throughout a brief suborbital flight, in order to prove that astronauts would be able to perform similar tasks satisfactorily.

There were three lights, with three levers directly below them on the device. One light was a red "continuous avoidance" signal which glowed all the time. Another was a white light that would illuminate when the test animal pushed the lever below. If they didn't do this every twenty seconds a mild electric shock flowed through metal plates attached to the soles of their feet. The third light was blue, and it would glow for five seconds at irregular periods every two minutes. The lever beneath this had to be pushed before the light went out or the chimp would receive a light shock. On an actual mission, this test was set up to begin at liftoff and continue through the flight, transcending periods of high g-loads and acceleration, weightlessness, and reentry.

In the post-flight *Review of Biomedical Systems for MR-3 Flight*, it was noted by Stanley C. White, M.D., Chief of the Life Systems Division, Richard S. Johnston, his assistant, and Gerard J. Pesman of the Crew Equipment Branch of the Life Systems Division, that the chimpanzee program was designed to parallel that of the human program.

"Its primary goal was the qualification of the man support systems," the report said. "Through this approach, the objective of flying first unmanned, followed by an animal flight, would give the logical sequence for the qualification of the spacecraft for manned flight.

36 The Mercury flight of chimpanzee Ham

Dressed for space, Ham demonstrates to his handler that he is ready to be considered for the MR-2 mission. (Photo: NASA)

Flight training for the chimpanzees involved learning to push levers in sequence with cueing illumination. (Photo: USAF)

Training for space 37

Ham, strapped into his couch and fully enclosed within his space container. Note the psychomotor panel and levers in front of him. (Photo: NASA)

"The chimpanzees considered for the Redstone program were thoroughly trained using the calculated flight dynamics. The centrifuge and heat chambers were used. The physiological training was incorporated with the psychomotor tasks to be done by the chimpanzee during flight. It was found that early in the training program the chimpanzee would cease working during the accelerative periods, and assume his normal

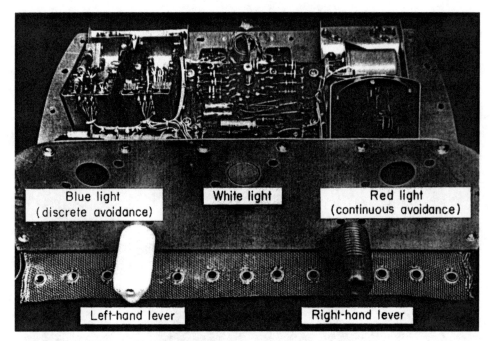

The MR-2 psychomotor panel. (Photo: NASA)

trained pattern promptly after the forces were released. However, subsequent training indicated that the chimpanzee could accept these new stresses and continue performance at a high level through all normal stress loads." [5]

Throughout the chimpanzees' training, a corps of veterinarians closely monitored their health and well-being, tracking their skeletal development with periodic exams and X-rays, as well as ensuring that they were free of any parasites. The animals also received regular checkups of their heart and muscular reflexes. Diet and dietary supplements were an important aspect of these tests, so the animals were fed small doses of antibiotics stirred into their favorite treat – liquid raspberry gelatin. In fact some of the primates enjoyed the diet and attention so much that they began to pack on excess weight, eventually washing them out of the program when they exceeded the specified limit of fifty pounds.

Even though Ham/Subject 65 trained well and was fast becoming one of the top candidates for the MR-2 shot, there were many physical, stress and readiness factors involved in the final selection – which was to be made on the eve of the mission. In preparation for MR-2, six of the most promising candidates along with 20 Holloman scientists and technical personnel were flown to Cape Canaveral on 2 January 1961 in order to acclimatize the chimps to a change in environment and to undergo final preparation for the flight, scheduled for the end of that month. Here they would be given 29 days of intense training under the supervision of Maj. Dan Mosely, DVM, in charge of Holloman's vast Aeronautical Branch.

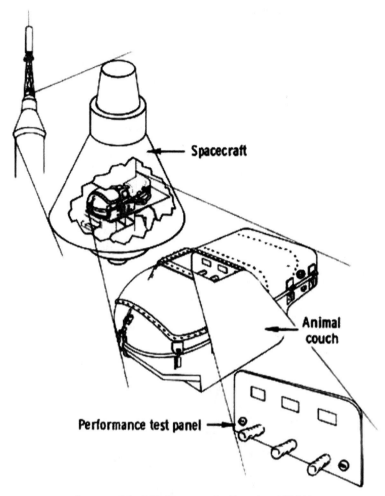
Layout of the MR-2 spacecraft. (Drawing: NASA)

Facilities at the Cape for quartering, training, and preparing the six chimpanzees consisted of seven specially designed trailers in a fenced-off enclosure adjacent to Hangar S, in which the astronauts' quarters were situated. To prevent any possible spread of disease amongst the animals they were isolated in separate cages. One of the trailers was a combined clinical and surgical facility for physical examinations, clinical laboratory analysis, minor surgery, and treatment of illness or injury. It was also used for the installation of biosensors, donning the restraint garment, and the placement of each chimpanzee in its personalized couch.

According to a report on MR-2 operations compiled post-flight by Capt. Norman Stringely and Maj. Mosely of the Air Force, and Charles Wheelwright from NASA,

A helmeted Ham in the lower section of his couch container. (Photo: NASA)

"Five practice countdowns were conducted by the medical preparation team for the MR-2 flight. They consisted of preparing the subject and couch, and proceeding up the gantry. The couch was either placed outside or inserted into the spacecraft and connected to the spacecraft environmental control system and electrical system. One countdown was for a telemetry check, one for a spacecraft-pressure check, one for a radio-frequency compatibility test, and two were simulated flights." [6]

PRELUDE TO FLIGHT

Three days prior to the launch of MR-2, newspaper reports across the United States were abuzz with a mounting air of excitement and expectation, as the flight of the chimpanzee was rightly being viewed as a prelude to the first flight of a human into space. An Associated Press report on Friday, 28 January 1961, described the build-up to the mission at Cape Canaveral, stressing that good visibility at the launch site and crucial points down the Atlantic test range was an essential requirement for the liftoff to proceed:

> In another 24 hours, if there are no delays, scientists will take a final look at six chimpanzees in their quarters here and pick one for the honor of being the nation's first animal astronaut to check out a Mercury spacecraft like those human astronauts will ride in later launchings. Then, six hours before launch time, the chimp will be packed into its own special space couch in a pressure chamber inside the nine-foot-high Mercury capsule.
>
> If the shot goes, this chimp – a mild-looking member of a specially trained team of four females and two males – will discover for science in a space of 16 minutes whether an animal, much like man in many ways, can tolerate the fantastic stresses of rocket flight under conditions of weightlessness in airless space.
>
> The launching vehicle will be a special Redstone missile which will hurl the chimp 115 miles up and 290 miles downrange at a speed hitting a peak of 4,000 miles an hour.
>
> If chimp and spacecraft make the flight okay, a human astronaut will try it in the next three months. Then, if a host of other trials go well, another chimp will be fired into orbit, and another astronaut will follow his trail, late this year or early next [7].

At 8:00 p.m. (EST) on Monday night, James Henry and John Mosely were on hand to select the prime and backup candidates. According to Ed Dittmer, "We didn't know which chimp would be going until the day before launch. There were six of them that were selected and they were all good, but Ham easily stood out as the best of the bunch." [8] Henry and Mosely agreed with Dittmer's judgment, selecting Ham because of his solid work under test conditions, as well as his general good nature, physical well-being, and alertness at the time. He was also declared to be the best prepared of the six finalists, having amassed 219 hours of training over a 15-month period, including being subjected to simulated Redstone launch profiles on the centrifuge at the Air Force Aerospace Medical Laboratory in Dayton, Ohio. The preferred backup chimpanzee, one of the four females, was Subject 46, known to her handlers as Minnie. She would be prepared to replace Ham at short notice should he develop any late abnormalities.

GOOD TO GO

In the very early hours of Tuesday morning, 31 January, Ham and Minnie were given a final physical examination. At 1:45 a.m., having been fitted with biomedical recording sensors and dressed in disposable diapers and plastic waterproof pants, both animals

Ham seated in his couch with backup Minnie looking on. (Photo: NASA)

waited patiently as an operational test was conducted of the sensors. They were then dressed in their spacesuits and firmly zipped and strapped into their individual contour couches. Psychomotor stimulus plates were then attached to the soles of each animal's feet and electrically checked for continuity. Their arms were left free in order that the one who flew the mission would be able to undertake the assigned psychomotor tasks aboard the spacecraft.

As preparations continued around them just after 3:00 a.m., Ham and Minnie enjoyed a prescribed breakfast consisting of some cooking oil and flavored gelatin, half a fresh egg, half a cup of baby cereal, and several spoons of condensed milk. All the tests had determined that Ham remained the better behaved and more animated of the two chimpanzees, and his place in space flight history seemed assured.

The next step in the proceedings was to install and bolt down the lids covering the chimpanzees, following which inlet and outlet air hoses were fitted and the air flow initiated. The containers were then checked for any air leakage, but all proved to be in order. At 5:04 a.m., after all the pre-flight tests had been satisfactorily completed, the handlers were instructed to drive the transfer van over to the launch pad, arriving 25 minutes later. Once there, Ham's container was switched from the transfer van's air supply to a portable oxygen supply, then carried to the gantry and up the elevator to the spacecraft level. After being gently inserted and secured into the capsule it was connected to the onboard environmental control system and electrical system. The physiological monitoring of Ham was then switched over to the blockhouse. Hatch closure was completed at 7:10, with an anticipated liftoff time of 9:30 a.m.

Ham's container is carefully inserted into the Mercury spacecraft. (Photo: NASA)

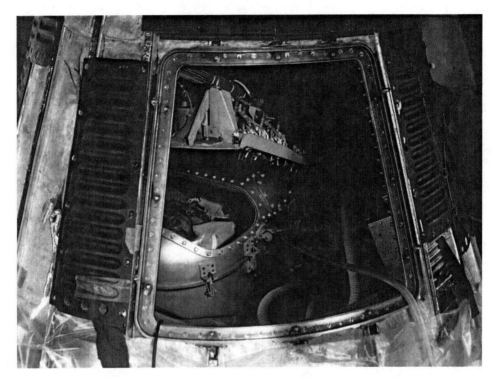

The interior of the spacecraft prior to hatch closure. (Photo: NASA)

Before the gantry was removed from the Redstone rocket at 8:05 a.m., the transfer van, with the fully prepared backup chimp still aboard, was moved a safe distance away, adjacent to the blockhouse. Still enclosed in her container, Minnie would be monitored up until 30 minutes prior to liftoff, at which time the container with its portable air supply and all her attending personnel would exit the van and move into the blockhouse.

At 9:08 a.m., the count was recycled and the gantry rolled back along its tracks into position. The spacecraft hatch was then opened to cool an overheated electronic inverter which was causing the temperature in Ham's container to rise. Technicians worked frantically to clear up a number of minor difficulties as concerns grew over a band of stormy weather closing in on the Cape. Repairs were soon completed and the countdown resumed at 10:15 a.m. But as the pad crew were evacuating the gantry its elevator jammed and had to be hurriedly fixed.

Liftoff finally occurred at 11:54:51 a.m. By then, Ham had spent nearly five hours strapped on his back inside the spacecraft. Two Mercury astronauts observed the ascent from the air, with Deke Slayton and Wally Schirra circling the launch area in F-106 jets.

Liftoff of the MR-2 mission. (Photo: NASA)

A TROUBLED FLIGHT INTO SPACE

The Redstone roared into the sky on what started out as the planned trajectory, but flight telemetry indicators soon began to show problems. A faulty valve was causing the fuel pump to inject too much liquid oxygen into the engine, inducing it to deliver an excess of thrust and accelerate faster than expected. As a result, the Redstone did more than was expected of it and, by burning its fuel faster than expected, triggered a chain of events which added several miles to the intended peak altitude and tacked 130 miles on to the range. Meanwhile, Ham was calmly pulling away at the levers as he had been trained to do.

When the booster exhausted its fuel supply, the Mercury spacecraft was meant to sequentially separate and coast to a peak altitude of 115 miles before falling into the Atlantic some 298 miles downrange, where a flotilla of eight ships were waiting to retrieve it. But the anomaly had caused a "thrust decay" when the rocket's fuel was depleted. That caused the spacecraft's emergency escape system to trigger an abort sequence. By then, the spacecraft was traveling at around 4,000 miles an hour. The emergency escape rocket reacted as it was meant to do, hauling the spacecraft away from the booster. In doing so, it accelerated to a speed of more than 5,000 miles an hour. Ham was suddenly subjected to a gravitational force of around 17 g's, driving him hard into his couch and making him temporarily forget his psychomotor duties. As the spacecraft finally entered a state of weightlessness a couple of small electrical jolts through the soles of his feet reminded a bewildered Ham of his responsibilities and he resumed tugging at the levers. But there were still more dangers to overcome.

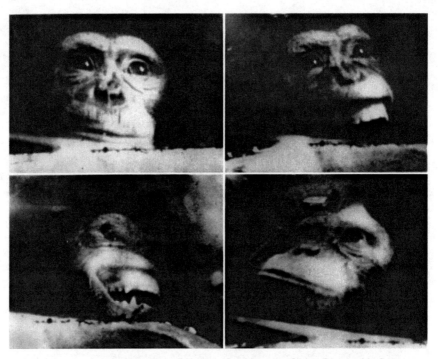

Still images from a film taken of Ham during his space flight. (Photos: NASA)

As Flight Director Chris Kraft and his Mercury Control Center team continued to monitor the progress of MR-2, he was informed that the fuel problem and resultant over-acceleration might carry the spacecraft an extra 42 miles higher and about 124 miles further downrange, adding two more minutes of weightlessness to the mission. Of more immediate concern to Kraft was the fact that a faulty relief valve had caused the spacecraft's pressure to suddenly drop from 5.5 to 1 psi. Fortunately, this would not affect the occupant, as Ham was sealed in a pressurized container with his own air supply. Added to this was the unhappy fact that the retro-pack had prematurely jettisoned when the spent escape tower was jettisoned. Consequently, the spacecraft would reenter excessively fast and splash down even further downrange.

William Augerson, a physician on duty in the Cape blockhouse, was monitoring Ham's physiological progress. He reported that despite all the onboard dramas, Ham was performing his tasks just as he had been trained. Weightless for more than six minutes, he only received two small electric shocks throughout the entire journey for neglecting to push the correct levers on time. In this respect, it was an almost perfect rehearsal for a manned mission, proving that a human would easily be able to carry out maneuvering tasks even if things did not go according to plan during the flight.

As MR-2 plunged backwards toward the sea, Ham began to experience a crushing 14.7 g's. Then, at 21,000 feet, a six-foot drogue chute automatically deployed, which in turn dragged the 63-foot main parachute from its stowage at 10,000 feet, rapidly slowing the spacecraft's rate of descent. A search and rescue and homing (SARAH) beacon had been activated earlier, when the escape tower pulled the capsule off the spent booster. Tracking aircraft monitored this signal and steered the ships of Task Force 140 to the predicted point of impact, around 416 statute miles downrange – an error of some 127 miles.

Seventeen minutes after lifting off, the capsule smacked down hard in rough seas beyond the far end of the Atlantic Missile Range. As intended, the landing bag had deployed and this helped to minimize the shock of striking the water. Immediately after splashdown the main parachute was automatically jettisoned, fluorescent green dye was released in order to aid visual sighting, and a high-intensity light on top of the capsule began to flash.

On impact with the water, a rim of the lowered heat shield had snapped back so violently onto the hull that it breached the titanium pressure bulkhead in two places, enabling sea water to penetrate the spacecraft. A cabin relief valve had also jammed open, allowing even more water to seep in. Then, just to compound matters, the heat shield tore loose from the bottom of the landing bag and sank. MR-2 slowly began to tilt and settle ever deeper into the tumultuous seas.

Shortly after splashdown, NASA was reporting that the floating capsule would be recovered within three hours. Although telemetry indicated that Ham was alive as the capsule approached splashdown, the radio telemetry circuits were disabled on impact so no one knew how he was doing. A subsequent NASA bulletin stated, "The Mercury spacecraft in today's test reached a velocity of more than 5,000 miles an hour, a peak altitude of about 155 statute miles, and landed some 420 statute miles downrange. Higher than normal booster thrust produced the extra velocity, altitude, and range.

The capsule has been sighted in the water by an aircraft. A recovery ship should reach the spacecraft within three hours. Telemetry received during the flight indicates the chimp performed satisfactorily." [9]

SPACECRAFT RECOVERY

Meanwhile the landing ship dock USS *Donner* (LSD-20), which had previously been involved in Mercury-Redstone recovery trials, was proceeding at flank speed to the reported landing area, together with Task Force destroyers USS *Ellison*, *Borie*, and *Manley*. Twenty-seven minutes after splashdown, airman technician Jerry Bilderback aboard a Navy P2V Neptune patrol plane became the first person to spot the capsule pitching around in white-capped seas. Unfortunately, the overshoot meant that the *Donner* was still some 60 miles away and it was almost an hour before the helicopter dispatched by the ship with pilots John Hellriegel and George Cox was able to reach the scene.

Once they were hovering overhead, the pilots alarmingly reported that the capsule was tilted on its side in a seven-foot swell, and it appeared to be sitting much deeper than expected in the water. By now, the destroyer USS *Ellison* had reached the site. With no time to spare, two trained frogmen quickly jumped out of the helicopter and attached cables to fixed points on the wallowing spacecraft to help keep it upright in the water. As the helicopter hovered, Cox reached down from the lower cabin with a shepherd's hook and attached a towline from the aircraft to a loop on the capsule.

At 2:52 p.m. Hellriegel applied full power and slowly hoisted the MR-2 capsule, streaming seawater, into the air. The precious cargo was flown all the way back to the USS *Donner* and gently deposited onto the deck at 3:38 p.m., where willing hands soon secured it. This good news was relayed to Cape Canaveral nearly three hours after liftoff.

When it was safe to do so, the spacecraft's steel hatch was removed, exposing the canister with Ham inside. The sailors involved also noticed a foot and a half of salt water sloshing around inside the capsule. It was later estimated the spacecraft had taken on about 800 pounds of sea water, but was otherwise in good shape. Happily, the water had not infiltrated Ham's container. He was unaware of how close he had come to sinking ignominiously to the bottom of the Atlantic.

Meanwhile, doctors back at the Cape were deeply concerned that Ham might have been injured during the crushing forces of the flight, or through the hard splashdown. About 35 minutes after reaching the ship, Ham's container was resting on the deck. One very confused chimpanzee could be heard squealing his discontent from within. The window was fogged over, but it cleared when oxygen was fed in through a small hatch, and Ham came into view.

"He's alive," reported a relieved Maj. Richard Benson, an Air Force veterinary doctor. "He's talking to us." The sailors then opened a small porthole to enable the veterinarian to insert his hand. Ham cried steadily. "That could mean some anxiety," Benson told the surrounding sailors. "He's just vocalizing."

Ham's spacecraft (circled at top) with the recovery helicopter overhead. At bottom (also circled) are two men in a raft near the bow of the USS *Ellison*. Their task was to right the capsule and help to attach a tow line so that it could be hoisted out of the water. (Photo: U.S. Navy)

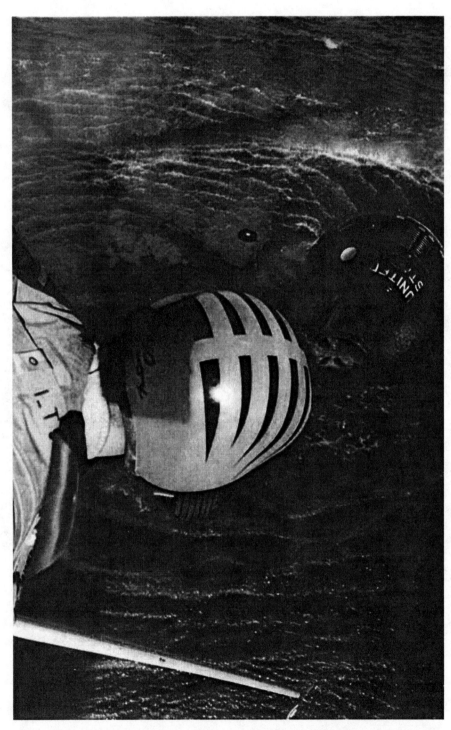

George Cox prepares to hook onto the wallowing spacecraft. (Photo: NASA)

Ham's spacecraft arriving by helicopter above the USS *Donner*. (Photo: U.S. Navy)

One sailor who got a glimpse of the animal was asked, "How does he look?"

"Fine," replied the sailor. "He's smiling at me."

Ham was turning his head from side to side, watching the onlookers curiously and licking his pink chops. He reached a couple of the fingers of his right hand through the port to grasp the hand of Benson. Then he rubbed his face and eyes and yawned. When the Plexiglas lid had been fully removed from the container, he once again shook hands with Benson, burped, and folded his arms across his chest while the veterinarian checked his heart rate with a stethoscope. Benson then reached down to test the animal's diapers. "They're damp," he said with a smile.

Following the brief checkup, Benson happily announced, "On the basis of this preliminary examination I'd say he looks very good. It is very encouraging." [10]

Ham was carried to the ship's battle dressing station and placed on a white table, where he was carefully unstrapped from his couch. Once again Benson checked the chimpanzee's heart rate, as well as his temperature, respiration, and lung conditions, and looked for any evidence of broken bones. Unsurprisingly, Ham did display some signs of fatigue, a little wobbling and trembling of his legs when standing, and he had somehow sustained a slight abrasion to the bridge of his nose.

Apart from the facial abrasion everything was fine, and Ham's reflexes were also found to be normal. Benson then produced a shiny red apple, at which Ham became excited, jumping and reaching out in anticipation. Benson cut the apple and fed it to him in slices as a post-flight treat, which he eagerly devoured. The flight had clearly

52 The Mercury flight of chimpanzee Ham

Pilot John Hellriegel gently lowers the MR-2 capsule onto a platform. (Photo: U.S. Navy)

Opening the hatch on Ham's capsule. (Photo: NASA)

not affected Ham's appetite. While he ate, Ham stood with his arm around the major, and later consumed half an orange along with a small wedge of lettuce.

Later, with Benson sleeping in an adjoining stateroom, Ham spent the night in the commodore's quarters as the ship steamed across a moonlit ocean for Grand Bahama Island. It was not exactly a trip of luxury, because he was in a cage on the floor of the bathroom, lashed to the toilet and the safety rail that was designed to prevent one from slipping after a shower aboard a rolling, pitching ship. But these were merely safety precautions aimed at protecting precious government research property [11].

UNWANTED FAME

The next day, Ham was loaded back onto the helicopter and transported to a forward medical facility at Grand Bahama Island for further medical checks. Once these were done, he was flown back to Cape Canaveral aboard a U.S Air Force C-131 transport aircraft, touching down at Patrick AFB at 1:11 p.m., where hordes of reporters and photographers were eagerly waiting alongside Hangar S for a glimpse of America's latest space hero.

Ham's container after extraction from the spacecraft. (Photo: NASA)

Ham was quick to indicate his displeasure at this rowdy intrusion into his living space. He became agitated, bared his teeth, and screeched at the melee of strangers. His handlers finally took the fretting animal back into the familiar surroundings of his van to calm him down. Upon being taken out again a short while later, he threw another tantrum as the news crews surged in close, some popping flashbulbs in his face. The handlers tried hard to get the reluctant chimp to pose next to a Mercury training capsule, but he didn't want to go anywhere near the darned thing. America's astrochimp was definitely not impressed by his newfound fame [12].

Several days later, on 3 February, Ham was returned to Holloman AFB in New Mexico. Here, over the next two years, he was kept under scrutiny while performing tasks to determine whether he had suffered any residual effects from his journey into space.

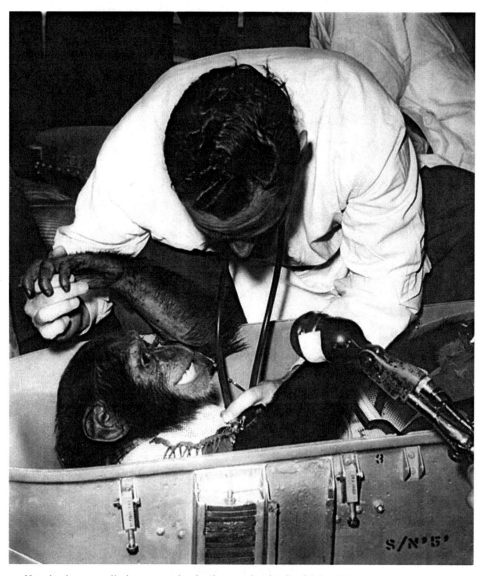

Ham is given a preliminary examination by veterinarian Dr. Richard Benson. (Photo: NASA)

Although he did train for a second mission, Ham never flew into space again. He spent 17 years in "retirement" at the National Zoo in Washington, D.C. In 1980, by now seriously overweight, he was transferred to the North Carolina Zoological Park, where he died as a result of an enlarged heart and liver failure on the afternoon of 17 January 1983, aged 26. His skeleton would be retained for ongoing examination, but his other remains were buried in a place of honor with a carved marker and memorial plaque outside of the International Space Hall of Fame in Alamogordo, New Mexico.

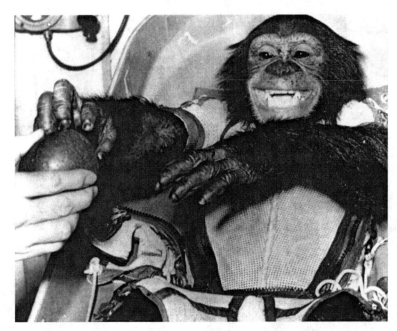

Ham eagerly reaches out to take an apple from Dr. Benson. (Photo: NASA)

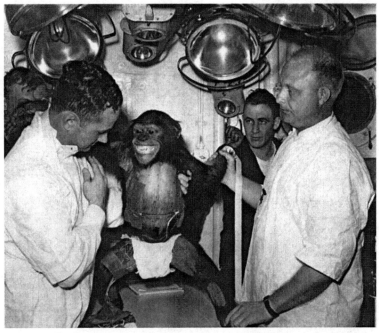

Dr. Benson (left) and M/Sgt. Paul Crispen remove Ham's biomedical sensors after his flight into space. (Photo: NASA)

The grave of space pioneer Ham in New Mexico. (Photo: International Space Hall of Fame, New Mexico)

The author stands alongside the MR-2 spacecraft, now on exhibition at the California Science Center, Los Angeles. (Photo: Francis French)

The positioning of the animal container inside the MR-2 spacecraft. (Photo: Francis French)

FINAL CHECKOUT OF THE REDSTONE BOOSTER

Following an extensive evaluation of the MR-2 Redstone's over-acceleration and harmonic vibration problems, it was reported that the reliability factor of the booster was well below the level required for NASA to confidently launch an astronaut into space.

Although the first manned flight with Alan Shepard as the prime pilot had already been scheduled for launch on 24 March 1961, there was a distinct feeling of unease in Washington, D.C. The president's technical advisor on science issues, Jerome B. Wiesner, had recently been appointed to head the Science Advisory Committee and was advocating a far more cautious approach in what he perceived as something of a rush for NASA to launch an astronaut into space. Wiesner bluntly warned Kennedy that a dead astronaut would not do a lot for the young president's administration, and he argued for several more chimpanzee launches to iron out any possible problems prior to committing to a manned flight. The new NASA Administrator, James Webb, and the head of the STG, Robert Gilruth, were brought into the discussion, holding consultations early in February with key Mercury personnel. Owing to some minor technical issues with Ham's flight, and under pressure from the White House to be cautious, Wernher von Braun was advised there should be a delay in the first human-tended mission. Instead, an unmanned proving flight of the booster would take place on the date previously allocated to MR-3.

As eager as he was to proceed with the manned flight, von Braun readily agreed with Webb and Gilruth – in fact, he had already been actively pressing for a further test flight, a "booster development launch" as he called it, although he was aware that it would not be possible to completely eliminate all risk. It was agreed that if this test proved successful, the manned MR-3 mission could proceed and the launch date was set for 25 April. It was a delay that arguably cost America the historical prestige of launching the first human being into space.

The new mission became known as the Mercury-Redstone Booster Development (MR-BD) flight. Its primary purpose was to verify the modifications made to prevent a recurrence of the flaws that afflicted the MR-2 flight. To prevent over-acceleration, the thrust regulator and velocity integrator were tweaked, and the vibration induced by aerodynamic stress in the upper part of the booster was remedied by adding four stiffeners to the ballast section and 210 pounds of insulation to the inner skin of the upper, instrument compartment section of the Redstone [13].

The MR-BD test would use an inert, expendable boilerplate Mercury spacecraft, and it was decided to reuse the one that had been retrieved after the Little Joe LJ-1B abort test mission on 21 January of that year. This capsule had been built at NASA's Langley Research Center, ballasted and configured to match the production capsule that was to be used on the first manned flight. However, it was not equipped with a retrorocket package or posigrade rockets because these would not be required. It was

The Manufacturing, Quality Control, and various other classifications of workers at the McDonnell Aircraft Corporation plant in St. Louis, Missouri, gather around the completed *Freedom 7* spacecraft, which would soon carry Alan Shepard into space. (Photo courtesy of Philip Kempland/McDonnell Aircraft Corporation)

Little Joe LJ-1B, launched on 21 January 1960. The boilerplate capsule used on this primate flight was recovered, and would later be used on the MR-BD flight. (Photo: NASA)

to be attached to the Redstone booster in the normal manner, but there would be no separation in flight. The escape rocket system, which was also inert, was a standard Mercury configuration utilizing spent rocket motors that were balanced to the correct weight for the MR-BD flight [14].

The LJ-1B flight had successfully carried Rhesus monkey Miss Sam on an eight-and-a-half minute test of the capsule's escape sequence and landing systems. The boilerplate capsule had splashed down smoothly 12 miles from the Wallops Island launch site on the Atlantic coast, whereupon it was plucked from the sea by a waiting helicopter and returned to the launch site. Forty-five minutes after liftoff, an excited but otherwise healthy Miss Sam was extracted from the capsule.

A SUCCESSFUL TEST FLIGHT

On 24 March 1961 the weather conditions at Cape Canaveral were favorable for a liftoff that day from Launch Complex 5. The launch procedures had been arranged in a four-hour countdown that began at around 8:30 a.m. (EST). Liftoff had originally

Rhesus monkey Miss Sam flew on the LJ-1B abort test flight from Wallops Island. (Photo: NASA)

been scheduled for 1:00 p.m., but this was advanced by half an hour at the request of the Atlantic Missile Range. The countdown would only involve procedures relative to the MR-5 Redstone booster, as the research and development capsule mounted on top was inert. Everything went smoothly, and the loading of the liquid oxygen began two hours prior to the scheduled launch time.

Including the spacecraft and its escape tower, the MR-BD vehicle stood 83.1 feet tall, and would have a total weight of 66,156 pounds at liftoff. Given the elongated fuel tank and enhanced performance of this Redstone variant, the more powerful but toxic Hydyne fuel was replaced by a mix of 75 percent ethyl alcohol and 25 percent water that would be combined, as previously, with liquid oxygen.

At 12:29:58 p.m. the MR-BD rocket lifted off the launch pad and booster cutoff occurred 141.7 seconds later. No thrust difficulties were encountered as the Redstone climbed to an altitude of 115 miles, attaining a maximum velocity of 5,123 miles an hour. After a flight lasting 8 minutes 23 seconds the entire assembly plunged into the Atlantic 311 miles downrange – almost exactly as planned. The area of impact was only 1.7 miles longer than planned, and less than 3 miles to the right of the envisaged

The Mercury-Redstone Booster Development (MR-BD) test that was launched on 24 March 1961. (Photo: NASA)

site. As the structure sank swiftly to the floor of the ocean, a SOFAR (sound fixing and ranging) bomb attached to the interior of the capsule automatically detonated at 3,500 feet. This device had been inserted at the request of the Navy for a checkout of its Broad Ocean Area (BOA) Missile Impact System.

All of the test objectives of the MR-BD mission were achieved, and a preliminary analysis of the flight data showed only slight deviations from the ideal performance. All systems functioned as planned and no problem areas were revealed.

"The engine performed perfectly," Dr. Kurt Debus, NASA's director of launch operations later explained. "It burned its prescribed time and did not cut off too soon, as on the previous launching." Debus announced that if a careful analysis of all the post-flight data demonstrated that the Redstone had functioned smoothly, no further tests would be required and that an astronaut would be able to be launched within six weeks to fly approximately the same 15-minute course as had been traveled that day. "However," he cautioned, "a close look at the tapes might reveal a slight flaw which could necessitate another test Redstone launching." [15]

Other NASA officials warned against an over-optimistic timetable, emphasizing that a manned flight depended on several other factors. Mercury Operations Director Walter Williams, pointed out that, in particular, the capsule had to undergo further helicopter drop and flotation tests before an astronaut could ride it.

RUSSIA RESPONDS

The very next day, 25 March, the Soviet Union overshadowed the Redstone test by launching into orbit and recovering by parachute the *Korabl-Sputnik 5* spacecraft, which not only carried a small dog named Zvezdochka ("Little Star") but also a full-sized space-suited mannequin cosmonaut which had been gleefully nicknamed "Ivan Ivanovich."

Now suitably armed with a launch date for the first American astronaut, whose name had not yet been publicly revealed, the Soviet Union pressed ahead in an effort to completely upstage and diminish America's space plans.

References

1. Telephone interview conducted by Colin Burgess with Edward C. Dittmer, 21 June 2005
2. *The Airman: The Official Magazine of the U.S. Air Force*, published by Defense Media Activity, USAF Office of Public Affairs, Washington, DC, issue April 1962
3. Burgess, Colin and Chris Dubbs, *Animals in Space: From Research Rockets to the Space Shuttle*, Springer-Praxis Publishing Ltd., Chichester, UK, 2007
4. Telephone interview conducted by Colin Burgess with Edward C. Dittmer, 21 June 2005
5. White, Stanley C., M.D., Richard S. Johnston and Gerard J. Pesman, *Review of Biomedical Systems for MR-3 Flight*, extract from *Proceedings of a Conference on Results of the First U.S. Manned Suborbital Space Flight*, combined NASA/National

Institutes of Health/National Academy of Sciences report, Washington, D.C., 6 June 1961
6. Stingely, Norman E., John D. Mosely, DVM, and Charles D. Wheelwright, Part 3, *MR-2 Operations* from *Results of the Project Mercury Ballistic and Orbital Chimpanzee Flights*, NASA SP-39, Washington, DC, 1963
7. *Lodi News-Sentinel* (California) newspaper article, "Chimpanzee May Go Into Orbit," Tuesday, 28 January 1961, pg. 9
8. Telephone interview conducted by Colin Burgess with Edward C. Dittmer, 21 June 2005
9. *The Dispatch* (Lexington, NC) newspaper article "Chimp Given Rocket Ride Over Atlantic," issue Tuesday, 31 January 1961, pg. 1
10. *The Norwalk Hour* (Connecticut) newspaper, article "Famous Space Chimpanzee Plays Ham After Thrilling Rescue, issue Friday, 3 February 1961, pg. 4
11. *Ibid*
12. Burgess, Colin and Chris Dubbs, *Animals in Space: From Research Rockets to the Space Shuttle*, Springer-Praxis Publishing Ltd., Chichester, UK, 2007
13. Swenson, Loyd S., James M. Grimwood and Charles C. Alexander, *This New Ocean: A History of Project Mercury*, NASA SP-4201, Washington, DC, 1989
14. *Memorandum on Mercury-Redstone Booster Development Flight (MR-BD)*, Space Task Group, Patrick AFB, FL. 26 March 1961
15. *The Schenectady Gazette* (New York) newspaper article, "Space Man Eyed After Rocket Shot Succeeds," issue Saturday, 25 March 1961, Pg. 1

3

NASA's first space pilot

"All of the first seven astronauts were real national heroes; not only to young people growing up in the late '50s and early '60s but to our folks as well." Lifelong Derry, New Hampshire resident David Barka was circumspect in looking back over several decades and recalling those times. "I would be hard pressed to equate their special status to any national figure living today. Keep in mind that the enthusiasm over the first space flights was fueled by the Cold War with the Soviet Union that had school children hiding under their desks during periodic air raid drills. The Soviets had been first in space and the Space Race took on almost a life and death feel."

David Barka grew up in the small farming and factory mill town which would one day gain an enviable reputation as America struggled to send a human into space for the first time. Named after Londonderry in Northern Ireland, Derry was first settled by Scottish-Irish immigrants in 1719. Some 50 miles from Boston, Massachusetts up Route 93 (now named the Alan B. Shepard Highway), the town has been home to several notable identities, including Matthew Thornton, a signer of the Declaration of Independence. For a time, the acclaimed poet Robert Lee Frost farmed and taught there while he wrote some of his epic works. In 1961, however, the town of Derry became forever identified with the supremely confident naval aviation officer, test pilot and NASA astronaut, Alan Bartlett Shepard, Jr., whose name is immortalized in our history books as America's first man in space.

"At some point I learned that Alan Shepard, born and raised in Derry, had been chosen to be America's first man in space," Barka reflected. "His folks still lived in the same house about a mile from where I grew up. I was nine years old then, and I couldn't have been more proud and excited. We were all aware of the great danger these men faced in going into space, especially Shepard, who would be first, and we were also aware that several rockets had blown up on the launch pads during test flights. On the day of the launch, Mrs. Blunt, who was the first-through-third-grade teacher at the Derry Village School, wheeled a television set into our classroom – something reserved only for special events – and we watched with great excitement the successful 15-minute flight of the first American in space.

The door knocker on the front of the house was part of the fittings when the house was first built, and is still in use today. (Photo: David & Debi Barka)

"Alan Shepard returned to his home town about a year later to a huge parade and celebration. The *Derry News* said 100,000 people watched the parade, an enormous event for a town of six to seven thousand people. I still have pictures. My family owned the Barka Oil Company and my father proudly had pennants made which had 'Spacetown U.S.A.' under the company name." [1]

A recent Christmas photo of the snow-covered Shepard house in East Derry. (Photo: Upper Village Hall, Derry)

CHEERING THE PRIDE OF DERRY

Due to Shepard's ongoing post-flight training commitments with NASA, it wasn't until 9 June 1962 that the people of the Granite State were finally able to openly express their admiration for the famed astronaut. Proclaimed "Alan Shepard Day" by Governor Wesley Powell, the occasion was marked by a well-planned parade that began at the Shepard family home in Derry and continued through the town streets. Accompanying the official cars in the preceding motorcade were some 2,000 people, 19 bands and 20 colorful floats. The bands played and the flags fluttered amidst a profusion of bunting and paper streamers, and crowds roared their welcome at each stop as Shepard and his family waved from their open convertible. Governor Powell made it to Derry for the special day, and was obviously swept up in the excitement when he somewhat grandly overstated, "This is the greatest day in the history of the state!" The motorcade terminated later that day at the steps of the state Capitol in Concord.

As the parade slowly progressed along the main street of Derry, some Navy patrol planes roared overhead, buzzing the town as part of the celebration, leading Shepard to comment with a wry smile, "I understand there's a Navy flier here who tried that once years ago, and didn't get away with it." [2]

At one of the stops along the circuitous route to Concord, the Shepards witnessed the dedication of a flagpole erected in his honor at Grenier Field in Manchester. He recalled for the assembled gathering that in his youth he used to sweep out hangars at the field in return for flying lessons. "This," he said proudly, "is where my original interest and devotion to aviation had its beginnings."

The tumultuous occasion proved a great inspiration for young David Barka. "It was then that my father and I took on a project together to build a coaster. We lived on a hill and all the kids in the neighborhood built carts to coast down the hill. Mine was in the shape of a rocket that I named *Freedom 7* in honor of Shepard's flight. That coaster sat in my folks' basement for close to 40 years, and when my Dad passed away in 1999 I couldn't bring myself to throw it away. I brought it to the house where I then lived with my wife and three children, not knowing that in 2002 we would purchase the Shepard home. The *Freedom 7* coaster is still here."

Today, David and Debi Barka reside in the large white colonial house, custom-built in 1921 on a 4.2-acre lot at 64 East Derry Road, the former home of America's first astronaut. "We love this house both for its beautiful traditional architecture and for its special history," he told the author. "We have modernized it where necessary, but preserved unique features such as the door casing that marked Alan and Polly's height as they grew, and the amazing built-in organ that Alan's father played; he was the organist at the First Parish Church down the road.

"My wife and I treasure the special meaning that this house has in the history of our country, and especially our town, and are happy to be a small part of it." [3]

Some of the Barka family assembled in front of their historic home. From left: Joe and his wife Nicole Barka, David Barka, Nick Barka, Debi Barka, son-in-law Mike McGivern with his and wife Anissa's son Finnegan. (Photo: David & Debi Barka)

A FAMILY'S HISTORY

Alan Shepard was an eighth-generation New Englander who could trace his roots back to the *Mayflower* as a celebrated descendant of Richard Warren (c.1580-1628), one of the first sea-weary passengers to set foot upon the snow-encrusted shores of what is now called Cape Cod following the ship's arrival on 11 November 1620. Ten years previously, he had married Elizabeth Walker in Hertfordshire in England, but in seeking a better life for his struggling family he had traveled alone by ship to the New World. Once he had established himself on a parcel of land in Plymouth, his wife and children Mary, Ann, and Sarah sailed on the ship *Anne* to join him. He and Elizabeth would go on to have two sons named Nathaniel and Joseph.

Remarkably for the time, their children survived to adulthood, were married, and had large families. Consequently, a vast numbers of Americans can today trace their ancestry back to Richard Warren and the settlement of America. In addition to Alan Shepard, Warren's descendants include such notables as Presidents Ulysses S. Grant and Franklin D. Roosevelt, and even the Wright brothers [4].

Shepard's middle name comes from his grandmother, Annie Bartlett, who in 1887 married Frederick J. Shepard in her home town of Nottingham, New Hampshire. The couple built a large home on farmland in East Derry and had three sons: Frederick, Alan, and Henry. Born in 1891, Alan, who was better known as Bart, was the father of future astronaut, Alan B. Shepard, Jr.

EARLY INFLUENCES

The man who would become the first American to venture into space was born on the upper floor of the family home in East Derry, on what he described many years later as the "bright autumn day" of 18 November 1923.

Alan Bartlett Shepard, Jr., was the first child born to Renza and Alan Shepard, Sr. While his mother had been born in Mobile, Alabama as Pauline Renza Emerson, she always preferred to be known by her middle name. Similarly, two years later, they named their new baby daughter Pauline, although everyone called her Polly.

Alan grew up on what was then a sprawling, picturesque small-town family farm. Even at an early age he knew the meaning of discipline. He and his sister had certain chores to perform, and their parents insisted that they be done at the prescribed time, ahead of any leisure time. Renza Shepard would later state that pursuing an orderly schedule of work and play helped Alan to develop a sense of duty.

"Our family did so much together that one member of the family could always depend on the cooperation of the rest," Renza stressed. "A sense of patriotism was also important in our family, and it was instilled in our children at early ages. Our house was always full of Alan's friends," she added. "He was a happy-go-lucky boy, very easy to explain things to, and very cooperative. Oh, he got in the usual amount of mischief, I suppose, but never anything serious. But my, how active he was!" [5]

His father had been commissioned a first lieutenant in the Army and was based at Fort Devens, Massachusetts, later serving in France during the First World War. He was recalled to active duty in the Army during 1940, and became a colonel in the Army Reserves. Alan Shepard Sr. was a treasurer of the Derry Savings Bank, owned the Bartlett and Shepard Insurance Company, and served as an incorporator at the Amoskeag Savings Bank in nearby Manchester. For many years, he was a treasurer and trustee for the Pinkerton Academy and a member of the First Parish Church of Derry, where for many years he fulfilled the role of treasurer as well as being their long-time organist. Having begun playing the original pump organ there at the age of fourteen, he served as church organist for the next 60 years.

Young Alan obtained his early education at the nearby Adams public elementary school, formerly the Adams Female Academy, where even as a small boy he began to excel in mathematics. In an interview with the Academy of Achievement in 1991 he spoke with fondness about his first teacher, Bertha Wiggins, and the influence she had on him.

"She was about nine feet tall as I recall, and a very tough disciplinarian. Always had the ruler ready to whack the knuckles if somebody got out of hand. She ran a well-disciplined group. I think most of the youngsters responded to that. There were one or two that couldn't handle it, and obviously they dropped by the wayside. But that still sticks in my mind. That's the lady that taught me how to study, and really provided that kind of discipline, which is essentially still with me." [6]

Shepard once skipped a grade because he was doing so well and Bertha Wiggins decided he needed to have more of a scholastic challenge. Although he often found it difficult, he rose to that challenge. "He was a little less bouncy in the classroom after that," his mother reflected. After completing five years of study at Adams School, Shepard attended junior high at Derry's Oak Street School.

THE URGE TO FLY

In his youth, Shepard developed a fascination with flying. One momentous day the family flew to Boston from Grenier Field, an Army Air Force base in Manchester. From then on he began haunting the airfield, about eight miles from home, watching airplanes take off and land, and later doing small helpful jobs around the hangars. He made dozens of model planes and read every book on flying that he could find, his favorite being a well-thumbed copy of Charles Lindbergh's epic, *We*.

"He loved science, too," his mother recalls. "Oh, and he had a boy-like interest in Buck Rogers and some of the science fiction. But in school he developed a serious interest in the subject and worked hard at it. He did extremely well in mathematics, and this helped him immensely." [7]

After junior high, Shepard went on to get his secondary education at the Pinkerton Academy in Derry, where he would complete grades 9-12. Ivan Hackler, who taught Shepard, recalled that his student had a keen interest in science as a youngster. "A good student and a boy extremely well-liked," Hackler said. "He was a good athlete, particularly in baseball and football." [8]

A recent photo of Pinkerton Academy. (Photo courtesy of Brian Chirichiello)

Outside of school hours, Shepard would deliver newspapers on his bicycle and attend Sunday school at the East Derry church, where his father was the organist.

"I was raised, if not exactly in an atmosphere of aviation, at least in the midst of mechanical things," he revealed in the Mercury astronauts' book, *We Seven*. "I had a five-horsepower outboard motor which I used to take apart and put back together again. And I often helped my father when he had things to tinker with – as you usually do in a small farming town. When I was in high school, a friend of mine and I used to cycle out to the airport … and do odd jobs around the hangar in exchange for a chance to take rides in an airplane now and then." [9]

From East Derry he went to school for the next year at the Admiral Farragut Academy in Toms River, New Jersey, specifically to help prepare for enrollment at the U.S. Naval Academy at Annapolis. He took with him a letter from his Pinkerton history teacher which spoke of his "good abilities" and "qualities of leadership." The Farragut records reveal that Shepard had a genius level IQ of 145, as proved by his solid marks in geometry and mathematics, but he did tend to lag a little in English. Noting this, his father wrote to the academy, suggesting he would appreciate it if a little more pressure could be placed on his son to do more studying. This obviously

Alan Shepard second from left in back row, taken at Pinkerton Academy in 1938. (Photo: Shepard family)

worked, as Shepard passed the entrance exam the following year at Annapolis with a 3.7 in math and a 3.3 in English out of possible perfect scores of 4. The Farragut Academy year-book recorded of him, "he speaks words of truth and soberness." [10]

He entered the Naval Academy as a 17-year-old "plebe" in 1941 and was whisked through the school at the wartime-accelerated pace, graduating with his bachelor of science degree a year earlier than the normal four-year term in 1944. The Academy's Professor George Beneze remembers Shepard for having a keen interest in aviation, pursuing associated subjects such as internal combustion and thermodynamics with zeal. "He asked a lot of questions, and was interested in what was going on in the laboratory," Beneze said. "You could depend on him." Classmates of Shepard at the Academy recall him as being energetic and aggressive. Among other activities, he rowed on the varsity crew in the bow seat. The story is that the coaches wanted to move him from that position because he was too light, but Shepard refused to move, declaring, 'I want to be first.'" [11]

On 3 March 1945 he married Louise Brewer, whom he had met while attending the Naval Academy. They would eventually have two daughters Laura and Julie, and also raise a niece, Alice, as part of their family.

"The Navy had a rule that even prospective flyers had to go to sea first," Shepard reflected, "so I spent some time on a destroyer [USS *Cogswell* (DD-651)] in the Pacific during the closing days of World War II. I took flight training at the Navy schools at Corpus Christi and Pensacola, Florida. Then I served in a fighter squadron that was based at Norfolk and made two cruises aboard carriers in the Mediterranean during 1948 and 1949.

"My flying career really got going in 1950 when I was still a lieutenant, junior grade, and was lucky enough to be chosen to attend the Navy Test Pilot School at Patuxent River. This was a real plum, especially for a junior officer." [12]

Alan Shepard as a U.S. Naval Academy "plebe." (Photo: USNA)

His parents were delighted with their son's latest achievement, as his mother Renza declared in a 1962 interview. "After Alan went on to the Naval Academy and took up the career of his dreams – being a pilot – he still missed his lovely New Hampshire. Many times he'd fly home, if only for a day. Sometimes he wouldn't be able to stop and see us, but he would buzz the house in his Navy fighter and we'd know who it was!" [13]

After graduating from Test Pilot School, Shepard stayed on at the Patuxent River Naval Air Station for a further two years, testing and aiding in the development of a number of highly powered Navy aircraft, such as the F-3H Demon, F-8U Crusader, F-4D Skyray, and F-11F Tigercat. A skilled pilot, he took over as Project Test Pilot on the F-5D Skylancer.

Among his many achievements as a Navy test pilot, Shepard says he "helped to develop the Navy's in-flight refueling system and was involved in testing the first angled deck on a U.S. Navy carrier … I was operations officer for a while in a night-fighter squadron that operated off the West Coast, and served aboard a carrier in the Pacific from 1953 to 1956." [14]

In 1958, following more flight-test and instruction work at Patuxent River, he was assigned to the Naval War College of Newport, Rhode Island in order to brush up on a number of academic subjects. Next, Shepard joined the staff of the Commander in Chief, Atlantic Fleet, as aircraft readiness officer.

The following year, an opportunity came his way that he found hard to resist. He had read articles on NASA's space program in newspapers and knew that the space agency would soon be recruiting a cadre of test pilots that they had begun calling "astronauts", meaning voyagers to the stars. Shepard knew that aviation was coming to an inevitable crossroad and space flight was the way of the future. He felt it would give him the chance not only to be regarded as a top pilot, but also as a space pilot, or astronaut. He was certainly qualified from all that he had read, and was ready to be called.

"I assumed they'd probably be looking me up and asking me if I was interested. As it turned out, they were doing just that. But the orders got misplaced somewhere – they wound up on someone else's desk for a few days – and I was beginning to wonder if I had been overlooked or disqualified. The orders finally came asking me to report for the first briefings, and I was delighted. I had a long talk with my wife that night, discussing what I should do if I were selected. Finally, Louise said, 'Why are you asking me? You know you will do it, anyway.' Louise had always been in complete support of what I had done, and I knew she was behind me now." [15]

SELECTING THE SEVEN

On 2 April 1958, in response to Soviet space efforts that were proving demoralizing to the American public, President Eisenhower had sent a bill to Congress calling for the immediate establishment of a civilian aeronautics and space agency. Congress passed the Space Act on 29 July, resulting in the creation of NASA, which officially came into existence on 1 October.

A Space Task Group was formed at the Air Force's Langley Research Center, Virginia, on 5 November, with Robert Gilruth appointed as director. On behalf of NASA, this task group was given four major objectives: to prepare specifications for a manned spacecraft; to plan and build a world-wide tracking network; to select and develop a suitable launch vehicle; and to select and train potential space pilots who would undergo a two-year training program.

With no precedents or government procedures to follow, NASA had to decide where the best candidates could be found, how many were required, and how they should be tested. What they *did* know was that the astronaut selection process would hinge on three crucial factors: physical, psychological, and technical.

In the final week of 1958, after several meetings between NASA Administrator Keith Glennan and his deputy Hugh Dryden, Robert Gilruth, and other upper-level representatives of NASA and the Space Task Group, a consensus was reached. For speed and facility in arriving at the selections, it was decided to restrict the search to the ranks of military test pilots. There were several reasons for this: test pilots were familiar with the rigors of service life, they were available at short notice, and their full service and medical records were on file for scrutiny.

It was decided to carry out the medical testing at an independent medical facility in New Mexico called the Lovelace Clinic, and to conduct further stress testing and psychological evaluations at the Wright Air Development Center in Ohio, which had already been involved in evaluation testing of space candidates for other potential service programs.

The Space Task Group determined that any candidate had to possess a university degree; be a graduate of a test pilot school; have around 1,500 jet hours; be in superb condition, both mentally and physically; be no taller than 5 feet 11 inches – a height dictated by the confines of the Mercury spacecraft – and be less than 40 years of age at the time of selection.

The first task for those involved in the initial selection phase, or Phase One of the operation, was a trip to the Pentagon where they pulled and evaluated the records of 508 pilots against broad selection criteria, checked their medical records and reports by superior officers, verified that they had the minimum amount of jet flying hours, and assessed the type of flying involved. Out of 225 Air Force records screened, only 58 met the minimum requirements. Of 225 Navy records screened, only 47 made the grade. Of 23 Marine Corps records screened, only five met the minimum standards. Thirty-five Army records were screened, but none met the requirement of being a graduate from a test pilot school. Women were excluded from consideration as there were no female military test pilots. Hence out of the 508 records screened, 110 met the minimum standards.

Each of the 110 candidates was then ranked in terms of his overall qualifications and the reviews were then placed in ranking order, from the most promising to least promising. These men were to be brought to Washington under secret orders and in civilian clothing in order to be briefed at the Pentagon by a senior officer from their respective service, as well as NASA officials. The first two groups would each have 35 men, with the remaining 40 men forming the third group. The groups were to be briefed in successive weeks during Phase Two of the operation.

The first group of 35 candidates turned up at the Pentagon on Monday, 2 February 1959, where the Air Force candidates were initially briefed on Project Mercury and what it might mean for their service careers by the Chief of Staff of the Air Force, General Thomas White, while the Navy and Marine candidates were simultaneously briefed in another room by the Chief of Naval Operations, Admiral Arleigh Burke. Prior to this, the men knew very little of Project Mercury or what it entailed. After the service briefings, the men were gathered together in one room for a more specific NASA briefing and an outline of Project Mercury by Charles Donlan, the associate director of the STG, Warren North, a NASA test pilot and engineer, and Lt. Robert Voas, a Navy psychologist. The men were then told that if they wished to opt out of consideration at that stage it would not be held against them or noted in their service records.

Those candidates that were willing to continue to the next phase were subjected to a preliminary suitability interview by psychologists Dr. George Ruff and Dr. Edwin Levy, then they sat through a review of their medical history. Some men proved to be taller than the limit, and were eliminated from the process. After the second round of briefings the following week, a total of 69 men had been processed. Faced with a higher than expected volunteer rate, Donlan canceled the third group, since he had more than enough applicants to fill the intended twelve positions. Eventually, six of the 69 candidates were found to be too tall and 16 declined to continue, leaving 47. Further checks and testing by NASA eliminated another 15 candidates, bringing the number down to 32.

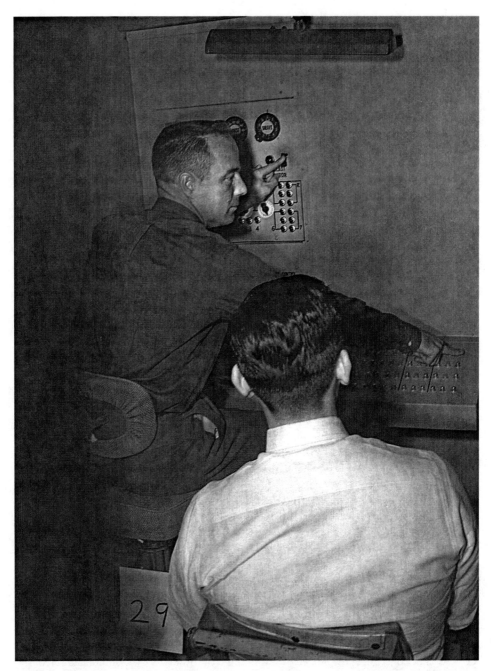

Mercury astronaut candidate Scott Carpenter undergoes reaction testing at the Wright Aeromedical Laboratory. (Photo: USAF)

All 32 men endured a meticulous, demeaning, and in some ways brutal week-long medical examination at the Lovelace Clinic in New Mexico. This was followed by another torturous week at the Wright Aeromedical Laboratory in Ohio, where they were subjected to extreme fitness and physiological testing, the purpose of which was to sort out the supermen from the near-supermen. Or to quote author Tom Wolfe on the subject, the selectors were seeking a group of men with "The Right Stuff."[1] In the process, one (James Lovell) was excluded for health reasons.

Then the results were compiled and considered, and the remaining 31 candidates were slotted into the following four categories:

Outstanding without reservations: 7
Outstanding with reservations: 3
Highly recommended: 13
Not recommended: 8 [16]

Early in March 1959 the results were forwarded to a panel at NASA Headquarters in Washington, D.C., for the final decisions to be made. It had been decided to halve the number required from twelve to six, but it proved impossible to decide between the final pair and so they were both accepted. Those chosen were notified on 2 April. A week later, on 9 April, seven test pilots were introduced to the waiting news media as the nation's first astronauts: Scott Carpenter, Gordon Cooper, John Glenn, Virgil ('Gus') Grissom, Wally Schirra, Alan Shepard, and Donald ('Deke') Slayton. They were about to be trained for a task beyond all others for a pilot – a flight into space with Project Mercury.

Giving his reason for wanting to become an astronaut, Shepard said, "I thought it was definitely a chance to serve my country. And I guess everyone feels an urge to do something no one else has ever done – the urge to pioneer and accept a challenge and try to meet it. I realized what it would mean to the Nation in prestige and morale. And I felt that I'd like to contribute whatever ability and maturity I had achieved. It would also, of course, be a big boost to my own self-confidence to know that I had done well in my chosen field. Every man needs that." [17]

TRAINING FOR SPACE

The seven astronauts were required to report for training at Langley Air Force Base on 27 April 1959. Here they were quartered in a two-story building dating back to World War I located on the east side of Langley Field, and given a very basic office that contained seven metal tables with name plates squatting on them, plus a young secretary named Nancy Lowe. If they had expected something a little more modern, given their new status in life, they were mistaken. On the other hand, they and their families were well used to the often Spartan lifestyle associated with being a military pilot.

---

[1] For a full description of the selection and candidate testing process, see the author's earlier publication, *Selecting the Mercury Seven: The Search for America's First Astronauts* (Springer-Praxis, 2011).

America's Mercury astronauts pose in alphabetic order front of a prototype capsule at the McDonnell Aircraft Corporation plant in Missouri. From left: Scott Carpenter, Gordon Cooper, John Glenn, Gus Grissom, Wally Schirra, Alan Shepard, and Deke Slayton. (Photo: NASA)

"We had started out at Langley Air Force Base in Virginia," Shepard said of their initial training. "NASA's new Space Task Group also set up its headquarters there. For the next two years we trained to be astronauts while engineers worked to develop and perfect the Mercury spacecraft." [18]

In the absence of prior ground rules for the training of astronauts, three basic philosophies were adopted: to employ any training device or method which offered even

Renza Shepard (left) and Alan Shepard, Sr., after the selection of their son as a Mercury astronaut in April 1959. Belle Emerson, Shepard's 90-year-old maternal grandmother, stands between his parents. (Photo: United Press International)

the remote possibility of being of value; to make the training with these devices as difficult as possible, even though analytical studies might suggest the task would be relatively easy; and to conduct the training on an informal basis because the men were all assumed to be well motivated, mature individuals.

They needed to brush up on basic mechanics and aerodynamics. As well, prior to being recruited they had been only briefly exposed to many fields of science such as astronomy, meteorology, astrophysics, geophysics, space trajectories, rocket engines, and physiology. As Mercury astronaut Deke Slayton would later report, "Instructors for these subjects were drawn from the scientists of the Langley Research Center and the Space Task Group. For example, one of the scientists of the Space Task Group gave us a lecture on the principles of rocket engines and rocket propulsion. Dr. William K. Douglas gave us a series of lectures on physiology designed to give us a better understanding of the physiology and construction of the human body, of which we had little prior knowledge. One of the subjects he discussed was the effect on the body of various g-loadings obtained during flight and landing impact." [19]

Bill Douglas was appointed as the personal physician to the group, and ended up going through many of the same training exercises and medical tests with the seven. To assist them in their training the STG brought in Bob Voas, a Navy psychologist, and Keith Lindell, a tough, regimented Air Force colonel. He would remain with the group until NASA transferred to Houston in 1962.

In addition to lectures on basic astronautics, the men were given detailed briefings by engineers from the McDonnell Aircraft Corporation on the design of the various subsystems of the spacecraft. And at formal briefings and coordination meetings the engineers within the Space Task Group responsible for the various subsystems kept the astronauts up to date with progress and the changes that were being made.

Supplementing their classroom or academic work, the seven also made several field trips as a group. One such excursion was to the Convair Astronautics Division of the General Dynamics Corporation in San Diego, California, for a tour of a test facility where components of the Atlas rocket, which was to supersede the Redstone and boost the Mercury spacecraft into orbit, were being tested. They also went to the McDonnell Aircraft Corporation to get their first look at a mock-up of the spacecraft, and observed the subsystems being manufactured. As a result of this visit, they made some recommendations for changes to the cockpit layout and instrument panel, and recommended the inclusion of a single large window and an explosive side hatch for escape (which would not be ready in time for the first manned mission). At the end of June they also went to the Redstone Arsenal at Huntsville, Alabama, where they observed Redstone launch vehicles being constructed and checked for flight.

"This was the home of Wernher von Braun. Wernher and his team weren't part of NASA yet," Slayton observed. "They still belonged to the U.S. Army, though it was clear they were anxious to move over. (The Army was eager for them to transfer.) Everybody was in a hurry. The Space Task Group was planning the first Redstone launches for early 1960, with orbital Atlas flights to follow in a few months. The whole program was supposed to be completed by the summer of 1961." [20]

A visit to the Rocketdyne plant followed, where the astronauts watched rocket engines being constructed and tested. As a group, they visited most of the facilities directly concerned with the launching of the Mercury spacecraft. In addition, they individually visited every subcontractor involved in the space program.

Since there was no way to replicate the weightlessness of space, the trainees flew in aircraft such as the C-131 on a succession of parabolic arcs. These simulations provided about 15 seconds of weightlessness as the airplane flew over the top of the maneuver. They also flew in the rear of a specially modified KC-135 (the military version of the Boeing 707) which gave approximately 30 seconds of weightlessness. The interior of the KC-135 was well padded, and they were allowed to float around at will. Weightlessness had no direct application to flight in the Mercury spacecraft, since the occupant would be securely strapped into a fairly small cockpit. By riding in the back seat of an F-100 at Edwards Air Force Base, California it was possible to achieve up to a minute of zero-g time and, remaining strapped in, they tested eating food and drinking water as they would in space.

As a follow-on to their zero-g training, the astronauts next underwent centrifuge training at the Naval Air Development Center in Johnsville, Pennsylvania. The 50-foot centrifuge there comprised a gondola mounted on the end of a large revolving arm, and it could simulate the dynamic loads of space flight from liftoff to reentry. A mock-up of the Mercury instrument panel had been installed inside the cabin of the gondola,

Gus Grissom (left) and Gordon Cooper during a simulated weightless flight in a C-131 aircraft at the Wright Air Development Center, Ohio. (Photo: NASA)

with the active flight instruments being driven by the centrifuge computer and a Mercury hand controller. It also held an environmental control system similar to that of a spacecraft. The gondola was sealed so that the men, clad in pressure suits and helmets, could depressurize it to the actual flight pressure of 5 pounds per square inch (psi). In this way, they could simulate flying at 217,000 feet with a 100-percent oxygen atmosphere at 5 psi, and could note the effects, if any, of applying high-g in the reduced pressure environment.

Each man made simulated flights both with and without the pressure suit inflated, and was able to rehearse all the normal launch, reentry, and abort profiles. The abort profiles could call for accelerations as high as 21 g, but they did not go to quite that level. Nevertheless, some of the astronauts underwent accelerations of 18 g without difficulty. The primary advantage of riding the centrifuge was the ability to evaluate techniques designed to help retain good vision and consciousness under high-g loads and to assist in breathing and speaking in such conditions. They would later say that the centrifuge was one of their most valuable training devices.

Gordon Cooper, Scott Carpenter and Alan Shepard at Johnsville, Pennsylvania for a week-long centrifuge training session in August 1959. (Photo: NASA)

A later NASA medical report observed that, "During the 12-month period prior to the flight, Shepard had completed three Redstone centrifuge training programs. He had undergone a total of 17 Redstone g-profiles in which he experienced cabin runs at sea level and at 5 pounds per square inch. These were rigorous programs, with emphasis on as accurate a mission simulation as possible. The astronauts used their personal contour couches, wore full pressure suits, breathed 100-percent oxygen, and performed a hand controller task. The electrocardiogram, respiration rate, and body temperature were recorded with each run, both static and dynamic. The runs were monitored by medical personnel using closed-circuit television from the centrifuge gondola, voice communication, and the physiological parameters noted previously. Physical examinations were conducted prior to and following the run sessions." [21]

Standing beneath the Johnsville centrifuge, Scott Carpenter looks up at Gordon Cooper and Alan Shepard on the gondola's access platform. (Photo: NASA)

Another complex test and training apparatus was the Multiple Axis Space Test Inertia Facility (MASTIF) located at the Lewis Research Center in Cleveland, Ohio. This device, made up of three metal frameworks – one within the other – featured a large seat mounted within the central gimbaled frame. Two feet in front of this seat were Mercury flight instruments, including a control handle which actuated a cluster of nitrogen thrusters. These high-powered jets were activated by test supervisors at an external control station, to set the MASTIF revolving simultaneously in all three axes at rates up to 30 r.p.m. The subject, strapped into the chair at the heart of a maze of aluminum spars, had to use the hand controller and the flight instruments to gain control over the yaw, pitch, and roll of the fiendish device and gradually restore its initial stable attitude. The men experienced very little difficulty insofar as the control task was concerned, but the multi-axis spin test did induce nausea after several such exercises.

Scott Carpenter and Alan Shepard beside the centrifuge gondola as Gordon Cooper prepares for a training run. (Photo: NASA)

Alan Shepard and Gus Grissom discuss training procedures while standing in front of the MASTIF device. (Photo: NASA)

Wally Schirra considered the MASTIF an inappropriate training aid. "It spun you up in all three axes at once – you were rolled, pitched, and yawed – to a velocity of 30 r.p.m. The seven of us rode the stupid thing, and we all lost our cookies or nearly did. Our main objection to MASTIF was that it didn't simulate a situation that was likely to occur in space – a spacecraft out of control in all three axes. The only time there was an incident the least bit like it was on Gemini 8, flown by Neil Armstrong and Dave Scott … and ironically they never trained on the MASTIF." [22]

Another valuable part of astronaut training was flying high-performance aircraft. Initially, they mainly flew a couple of F-102As, which were later exchanged for two F-106As. As they were brought into the space program as highly qualified jet pilots, the astronauts felt it highly desirable to maintain their basic proficiency.

Since the primary recovery mode was to splash down in the ocean, the astronauts were required to become proficient in water egress and survival training. Initially, an egress trainer was submerged in a hydrodynamics tank at Langley Research Center and they practiced egressing techniques, first in smooth water and then in artificially generated waves. When the men felt they had developed a reasonable proficiency in that facility, their capsule trainer was transported down to the Gulf of Mexico, near Pensacola, Florida, taken to sea on a barge and dropped over the side. The astronauts then practiced evacuation techniques in the open sea, which on many occasions was fairly rough.

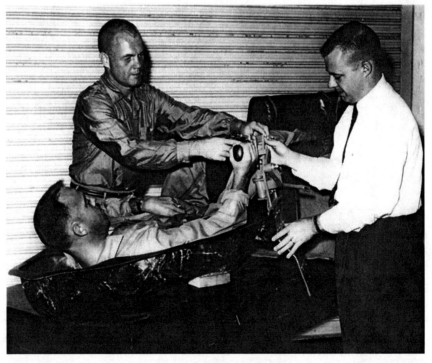

John Glenn and Alan Shepard, in contour couch, training with the spacecraft's hand controls under the watchful eye of Dr. Robert Voas of the Langley Field Research Center staff. (Photo: NASA)

One egress exercise involved the use of a helicopter hooking onto the spacecraft and lifting it partially out of the water, so that the lower frame of the door was above the water line. The astronaut climbed out of the hatch, then the personnel lifting line (dubbed the "horse collar") was lowered down to them and once they had donned it they were hoisted up into the helicopter. The advantage of this method of egress was that it proved to be the fastest means of leaving the spacecraft and being recovered by the helicopter. Alternatively, the astronaut could leave the hatch unopened and be carried inside the spacecraft to a waiting aircraft carrier. However, after a helicopter inadvertently dropped a spacecraft on the way to the recovery training area during one early exercise, the astronauts lost a lot of confidence in riding in the spacecraft in this manner.

In conjunction with the water egress training, the astronauts were given survival training, during which each spent half a day in a one-man raft learning how to distill water, to protect himself from the Sun, and to signal the rescue forces.

They also received training for the unlikely event of landing in a hot, remote area such as the West African desert. They spent three days at Stead Air Force Base, near Reno, Nevada, learning how to protect themselves from the Sun in that environment, how to utilize the limited water supply, and how to make clothing and shelter out of parachutes.

Supplementing all these training and survival techniques was a device known as the Flight Procedures Trainer. This was a basic mock-up of the Mercury capsule with its control systems connected to monitoring pick-ups and computers. It was able to simulate any conceivable routine or unusual performance of the spacecraft, allowing the astronauts to rehearse problem-solving techniques. One such trainer was installed in the Mercury Control Center at Cape Canaveral. This not only gave the astronauts a chance to react swiftly and safely to unplanned hazardous flight conditions during training, it also enabled the engineers that would monitor real missions to observe an astronaut directly controlling a Mercury spacecraft.

"We spent two years doing many things and following up many avenues to make sure that we had not overlooked anything," Shepard once pointed out. "We crammed ourselves full of knowledge. We built up our stamina on the big machines. And we got thoroughly familiar with the spacecraft that we'd fly. Some of this was fairly exotic stuff. For we were preparing to penetrate an environment that no one had ever dealt with before. Some of it, however, was just plain down-to-Earth hard work." [23]

SETTLING IN

With ever more of their training concentrated at Cape Canaveral, NASA eventually relocated the astronauts to crew quarters on the west side of the Cape, several miles from the Redstone launch complex. As at Langley, the quarters were nothing special; they were situated in an old concrete block Air Force hangar known as Hangar S, in which the Martin Company people had prepared the Vanguard rockets for launch. As with all such hangars at the Cape, its floor plan was very basic, having a two-story high-bay area in the center between two large sliding doors, and regular access doors at each end. Offices or stage rooms were located on both sides of each floor.

Dr. William Douglas, the astronauts' personal physician, was aided by a level-headed young Air Force nurse named Delores O'Hara, who would not only tend to any illnesses or injuries they or their families suffered, but eventually become a true and sympathetic confidante to the men and their wives. Dee O'Hara, as she is better known, was an Idaho-born 2nd lieutenant in the Air Force Nurse Corps serving at the nearby Patrick Air Force Base Hospital, when her commanding officer asked if she would like to be the nurse assigned to NASA's Mercury astronauts.

When the news leaked out of her assignment, the press began to hail Dee as the nation's "Space Nurse" and "Astronaut Nurse," but she insisted that the only proper designation was aerospace nurse.

The astronauts' nurse, Dee O'Hara. (Photo: NASA)

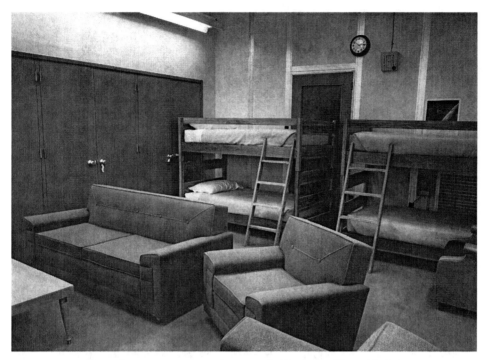

The rather austere astronauts' quarters in Hangar S. (Photo: NASA)

One of the first major responsibilities assigned to young Dee O'Hara was to help establish what was referred to as the Aeromed lab in Hangar S.

"I always felt the Mercury program was launched from Hangar S because we were all crammed in this one little hangar. The suit rooms were there, the capsule was there – everything existed in Hangar S. And so I set up the Aeromed lab. It was a long string of rooms. It was very narrow. We had a hallway, the rooms, and then it overlooked the floor of the hangar, if you will. I had a lab area, an exam room area, then there was my little office. Then we had a large room, a carpeted room where the suits were, the suit couch was there, and all of the suit checkouts were done there. And then you went to the next room, which was kind of set up as a conference room for them, and then past that was a little lounge area with a La-Z-Boy chair … it was considered crew quarters if you will. Primitive, but it was considered crew quarters. And in the next room, and these are all very small rooms, were bunk beds for if the astronauts stayed there. Most of them stayed in motels in Cocoa Beach, but if they were training late, or working late in the capsule, they could at least bunk there and sleep the night, and not have to drive. Because I think it was like nineteen miles or so back into Cocoa Beach. Granted, that's not a long drive, but you're out in the middle of nowhere; I mean of course it's all built up now, but back then it was not. I think the first motel bar was the Polaris Lounge or something like that, and so it was kind of distant to the Holiday Inn where most of them stayed." [24]

Despite their fame, the pioneering years of space flight were often rather less than glamorous for the Mercury astronauts. Shepard recalls the early ignominy of living alongside a colony of space-trained primates. Back then, the astronauts knew that they were being forced to play second fiddle to the chimpanzees. "The crew quarters at Hangar S were Spartan, austere, nondescript, and totally uncomfortable," Shepard wrote. "Our sleeping quarters could be reached only by going down a long, poorly lit hallway, an unpleasant walk during which we were assailed by the hoots, screeches and screams of a small colony of apes housed out back. In the end, we decided the humiliation of stepping aside for a monkey was bad enough. We certainly didn't have to live with the howling dung-flingers." As Dee O'Hara suggested, most of the astronauts moved into a Holiday Inn at Cocoa Beach [25].

Apart from their overall work on the Mercury program, each of the astronauts was handed an assignment which would become his specialized field. Each then shared vital information on his subject with the others. Scott Carpenter was given capsule communications and navigation; Gordon Cooper, engineering developments to adapt the Redstone missile to launch the manned capsule on suborbital flights; John Glenn was to monitor the layout of the capsule interior, as well as its instrumentation and controls; Gus Grissom was assigned the capsule's automatic pilot; Wally Schirra got the capsule's environmental control system; Deke Slayton looked ahead to mating the spacecraft with the Atlas booster for orbital flights; and Alan Shepard was given responsibility for the capsule recovery and rescue program.

McDonnell's Pad Leader, Guenter Wendt, was hard at work preparing for the first manned flight, but concerns were still cropping up that no one had considered, such as an effective means of evacuating an astronaut from a spacecraft in peril whilst on the launch pad.

"As we moved closer to our manned launches, it came to our attention that we had no means of egress for the astronaut if an emergency occurred while the launch tower was parked several hundred feet away. I think it was Bob Munger who came up with the idea of using the 'cherry picker.' This was a yellow, crane-like rig on a long flatbed truck. It was operated by remote control. Its long boom could remain in position close to the spacecraft until right before the launch. In an emergency such as a fire on the pad, the astronaut could open the hatch and climb into the metal cage on the boom's end. We would then swing him out of harm's way. It was a bit crude, but it did the job." [26]

"MY NAME, JOSÉ JIMÉNEZ"

If there was one person in his life that Alan Shepard truly came to admire after he was selected as an astronaut, it was a short, stocky fellow who was born in Quincy, Massachusetts with the tongue-twisted name of William Szathmary. Adopting the stage name of Bill Dana, the comedian became an instant hit with the astronauts, and in particular Shepard, due to his hilarious routines centered on a bumbling, nervous astronaut character named "José Jiménez."

Shepard could recite many of Dana's routines off by heart, and would often slip into the Jiménez character during training in order to relieve any pent-up anxieties. The astronauts soon got to know Dana well, and embraced him as one of their own. With their enthusiastic blessing he was forever endowed with the title of the "eighth astronaut."

Comedian Bill Dana as José Jiménez, the nervous astronaut. (Photo: Bill Dana)

Shepard once spoke about the origin of his close affinity with the much-loved comedian. "There was this TV program and I was crazy about it. It was so close to the way that we see things when we're in a good mood. In short, I liked him so much I recorded it on tape, then I took the tape to Cape Canaveral and during the Ranger launching, at a moment when they stopped the countdown because something was wrong, I put the tape on at full volume there in the control room. Sometimes we like to have a little fun too. It releases the tension."

As Wally Schirra recalled, "Dana was doing a one-night stand in Cocoa Beach near Cape Canaveral, and Al Shepard and I were in the audience. When Dana asked for a straight man, Al volunteered. Then I did, too. What we really liked about Dana is that he made himself the butt of his jokes. Like Bob Hope, also a dear friend of the astronauts, he did not get laughs at the expense of someone else." [27]

One skit which Shepard was especially fond of centered on an interview between a reporter (played straight by Don Hickley) and José Jiménez, chief astronaut of the Interplanetary Forces of the U.S.A. In part, the routine goes like this:

DH: The gentleman you're about to meet is the most important man in any of our lives. He's the United States' officer who has been sent into outer space. I'm referring to the chief astronaut with the United States Interplanetary Expeditionary Force, and here he is now. How do you do sir. May we have your name?

JJ: My name, José Jiménez.

DH: And you're the chief astronaut with the United States Interplanetary Expeditionary Force?

JJ: I am the chief astronaut, with the United States ... Interplanetary ... My name José Jiménez.

DH: Mr. Jiménez could you tell us a little about your space suit?

JJ: Yes, it's very uncomfortable.

DH: How much did the space suit cost?

JJ: That space suit cost 18,000 dollars.

DH: 18,000 dollars? That seems rather expensive.

JJ: Well it has two pair of pants ... so that's only 9,000 dollars apiece.

DH: I've been noticing this, Mr. Jiménez. What is this called? A crash helmet?

JJ: Oh, I hope not.

DH: I want to ask you: What is the most important thing in rocket travel?

JJ: To me, the most important thing in the rocket travel is the blast off.

DH: The blast off?

JJ: I always take a blast before I take off (pause) otherwise I wouldn't get in that thing.

DH: And I just wonder what you'll do to entertain yourself during those long, lonely, solitary, hours when you're all by yourself?

JJ: Well, I plan to cry a lot.

DH: I just wonder if there are a few words that you'd like to say to the people of the United States?

JJ: Yes, there are a few words that I'd like to say ...

DH: Please go ahead.

JJ: People of the United States of America ... please don't let them do this to me!

While there was often a mischievous, even impish side to Alan Shepard, he took the flying aspect of his new vocation very seriously. Those who knew him best often recall there were two sides to the man: there was Al who loved a good practical joke, and then there was the serious, pragmatic Commander Shepard who could cut people down with a withering stare when they did not meet the high expectations which he demanded in regard to their job functions. If he was in the latter mood, people only approached him with great trepidation.

"Al was the most complex of the original astronauts," Gordon Cooper revealed in his memoirs. "He seemed to have two distinct personalities: one the charming and beguiling jokester who introduced José Jiménez – comedian Bill Dana's popular alter ego – and his 'Please don't send me' astronaut act into our everyday lives; and the other which came out when the chips were down and was so competitive as to be ruthless. We all knew to watch our backs when *that* Al was around." [28]

DECISION DAY

Robert Gilruth was faced with a difficult decision, but one that he had to make in order to ramp up more specific mission-oriented training. As the head of the Space Task Group it fell to him to decide which of the seven astronauts would be assigned to attempt the nation's first manned space flight. Whilst he did have one person in mind, he felt it only fair to reinforce his opinion by running a selection poll amongst the men themselves. And so, in December 1960, they were all asked to vote for the person – excluding themselves – that they thought was the best overall candidate for this historic mission.

It would be a critical decision, as celebrated author Tom Wolfe wrote in his epic about the Mercury astronauts, *The Right Stuff*. "When they assembled in [Gilruth's] office, he told them he wanted them to take a little 'peer vote,' along the following lines: 'if you can't make the first flight yourself, which man do you think should make it?' Peer votes were not unknown in the military. They had been used among seniors at West Point and Annapolis for some time. For that matter, during the selection process for astronaut, the groups of finalists at Lovelace and Wright-Patterson took peer votes. But peer votes had never amounted to anything more than what they were *prima facie*: an indication of how men at the same level regarded one another, whether for reasons of professionalism or friendship or jealousy or whatever. Pilots regarded peer votes as a waste of time, because a man either had the right stuff in the air or he didn't, and a military career, particularly among those with 'the uncritical willingness to face danger,' was not a personality contest. But there was something about Bob Gilruth's deep concern … They were to think the whole thing over and drop them off at Gilruth's office." [29]

As Deke Slayton later recalled, "I think he just wanted to know if we agreed with his judgment."

For Shepard, teamwork was always secondary to his own ambition to be the first American astronaut to fly. "I knew there was a lot of talent there, and I knew it was going to be a tough fight to win the prize," he told interviewer Roy Neal. "It was an interesting situation, because I was friendly with several of them. And on the other hand … there was always a sense of a little bit of reservation, not being totally frank with each other, because there was this very strong sense of competition … seven guys going for that one job." [30]

They would not have to wait very long. The following month, on the afternoon of 19 January 1961, just a day before the presidential inauguration of John F. Kennedy, each of the seven received a phone call from Gilruth requiring them to assemble in their Langley office at five o'clock for an important meeting. Each man knew this was probably the day that one of their number would be handed that precious first flight.

An air of nervous tension filled the office as the astronauts fidgeted, waiting for Gilruth to arrive, trying to crack feeble jokes but with heart beat rates rising as five o'clock came and went. Finally, he entered the sparse office, closing the door behind him. Knowing that the seven men would be eager for his decision, he simply stood before them, cleared his throat, and spoke.

"Well, you know we've got to decide who's going to make the first flight, and I don't want to pinpoint publicly at this stage one individual," he began. "Within the organization I want everyone to know that we will designate the first flight and the second flight and the backup pilot, but beyond that we won't make any public decisions." Then he paused momentarily. "Shepard gets the first flight, Grissom gets the second flight, and Glenn is the backup for both of these two suborbital missions. Any questions?"

There was only a stunned silence in the room. No one dared to speak as they each digested the news. Gilruth paused a few moments, and as he turned to leave the room added, "Thank you very much. Good luck!"

Shepard recalls those few seconds after Gilruth left as a time of joy, triumph, and a wonderful sense of accomplishment, but he also knew he mustn't let this show. Six fine colleagues were coming to grips with the baffling news that they had not been considered the best candidate for the gold-ring job that they had all so badly wanted. With mixed emotions of elation and a sense of empathy coursing through him, it was tough simply being there. "I did not say anything for about twenty seconds or so," he later stated. "I just looked at the floor. When I looked up, everyone in the room was staring at me. I was excited and happy, of course; but it was not a moment to crow. Each of the other fellows had very much wanted to be first himself. And now, after almost two years of hard work and training, that chance was gone." [31]

The comradeship that would define the Mercury astronauts soon took over, as one by one the others moved over to congratulate Shepard, shaking his hand with smiles and encouraging words that could barely mask their dismay before leaving the room to come to terms with their bitter disappointment.

When he came home with the news that evening Shepard said to his wife, Louise, "Lady, you can't tell anyone, but you have your arms around the man who'll be first in space!"

"Who let a Russian in here?" she replied with a wide smile. It was a better joke than she knew [32].

STEPPING UP THE TRAINING

Much to the frustration of the media, waiting for news of the astronaut selected to make the first space flight that year, NASA would only announce that of the original seven, only three – John Glenn, Gus Grissom and Alan Shepard – had been selected as prime candidates for the first Mercury-Redstone mission, and the name of the man who would fly would not be revealed until nearer the time. The other two, as well as the remaining four men, would fly later space shots.

For Glenn and Grissom, and particularly Shepard, the training was stepped up to prepare them for that first, crucial flight. Shepard would find the pace particularly grueling at times.

"Early in 1961, the Cape, as it was simply called, was the most exciting place in the country. It was also a very tough place to work. Despite the glowing press reports about how well things were going with the astronauts and the Mercury operations team, the reality was that conflict was a part of everyday life at Cape Canaveral. Arduous project schedules and the long wait to get up into space made us feel stifled and resentful."

And then there was the matter of a primate making a suborbital flight ahead of an astronaut.

Until the flight was imminent the press and public would know only that the nation's first astronaut would be selected from John Glenn, Alan Shepard, and Gus Grissom. (Photo: NASA)

"The irony of playing second fiddle to a chimpanzee was particularly galling to us," Shepard noted. "NASA had decided to send a chimp into space before sending me. I protested again and again, but NASA insisted the little ape go first. The agency meant well. But all I could think about were Russian boosters rolling to their pads for the first manned space flight.

"There were other frictions too. At the Cape I spent most of my time in a 'procedures trainer.' This was a replica of the actual spaceship that would boost me more than one hundred miles into space. It also duplicated the severe semi-supine flight position, with the pilot lying on his back, legs vertical to the knees and then dropped down so that he was shaped like a squared-off pretzel. No one liked the trainer. It was like taking a straight-backed chair, placing it on its back, and then 'sitting' in it. This is where the astronaut trained to reach all his instruments and controls until he could go through every motion of his scheduled flight with his eyes closed and never miss hitting the right button or lever.

"By late January events were coming down to the wire. As flight time neared, the practical joking that had helped keep us all sane faded away. A serious tone settled over the launch, support, flight, and recovery teams. Redstone, the booster rocket, was working well, and I was scheduled to be launched in about six more weeks." [33]

CAUTION WINS, AMERICA LOSES

At 11:55 on the morning of 31 January 1961, chimpanzee Ham was launched on the suborbital MR-2 flight. The Redstone climbed steeply and then headed downrange over the Atlantic Missile Range to where a fleet of Navy vessels was stationed in the target zone some 290 miles from the Cape. Due to the spacecraft overshooting its intended splashdown area and Ham experiencing a rough reentry that exposed him to a hefty load of 18 g's, there were fears that the chimpanzee might not have survived. However, after the capsule was recovered, the attending veterinarian, Maj. Richard Benson, pronounced Ham to be "healthy and happy."

While there was relief that Ham's flight had ended well, Shepard was unshakable in his belief that a human astronaut should have occupied that capsule.

"I reviewed the telemetry tapes and records of the Great Chimp Adventure. I knew I could've survived that trip, but I also knew immediately that my own planned flight was in deep trouble. If only the damn chimp's ride had been on the mark, I'd have launched in March.

"But Ham's flight had not been on the mark, and in Huntsville, Alabama, Dr. Wernher von Braun, developer of the Redstone and director of the Marshall Space Flight Center, was showing signs of a new conservatism as responsibility for men's lives was factored into his decisions. 'We require another unmanned Mercury-Redstone flight,' he said. Working with the engineers, I confirmed that the problem with Ham's Redstone had been nothing more than a minor electrical relay. The fix was quick and easy, and the Redstone was back in perfect shape. 'For God's sake, let's fly. Now!' I begged NASA officials, but Dr. von Braun stood fast: 'Another test flight.' I stalked off

Redstone rocket No. MR7 that was to boost *Freedom 7* into space is prepared for transportation to Cape Canaveral. (Photo: NASA)

The MR-3 Redstone booster being raised onto the launch pedestal. (Photo: NASA)

steaming to the office of Flight Director Chris Kraft. 'Look, Chris, we're pilots,' I said. 'When there's a failure, dammit, we fix it.'

"'I know, Alan,' he said.
"'Well, what about it? It's an established fact that the relay was the problem, and it's fixed.'
"'Right.'
"'So why don't we go ahead? Why don't we man the next one?'
"'Why waste time, right?' Kraft smiled.
"'Right.'
"'Because when it comes to rockets' – the flight director shook his head – 'Wernher is king.'
"'King?'

Freedom 7 is hoisted up to be mated to Redstone rocket No. MR7 in preparation for the first flight by an American astronaut. (Photo: NASA)

"'King.'
"'Forget it, right?'
"'Right.'

"So I walked away, brooding. The March 24 Redstone flight was an absolute beauty. I could've killed. I should've been on that flight. I could've led the world into space. I should've been floating up there, while the Russians were still wrestling with a balky rocket booster."

By the time Shepard's flight was ready to go, Yuri Gagarin had already been there and back.

A publicity photo of Alan Shepard holding a model of a Mercury spacecraft and its escape tower. (Photo: NASA)

"So that was that," Shepard pointed out ruefully. "Nearly four years after *Sputnik* started the Space Race and two years after I and my six colleagues – Scott Carpenter, Gordon Cooper, John Glenn, Virgil ('Gus') Grissom, Wally Schirra, and Deke Slayton – were presented in a Washington, D.C., ceremony as the Mercury Project team that would represent America in space, we'd been beaten to the punch. We had them by the short hairs, and we gave it away." [34]

As Guenter Wendt reflected, "As we busied ourselves incorporating the latest changes into the spacecraft, Glenn, Grissom and Shepard stayed busy in the simulator in Hangar S. The three prime candidates for the first Mercury flight spent 50 to 60 hours a week working on procedures in the simulator. During that period, Shepard made about 120 simulated flights, some in the sim[ulator] and some in the altitude chamber. In spite of the fact that Gagarin had orbited the Earth, and our first flight would only be suborbital, our Mercury was much more sophisticated than their Vostok. Of course we didn't know it at the time, but the lead we had in spacecraft systems was one that we would never relinquish." [35]

References

1. E-mail correspondence with David and Debi Barka, 3 February – 11 March 2013
2. *The Victoria Advocate* newspaper (Texas), article, "Long-Delayed Tribute Paid to Alan Shepard, issue 10 June 1962, pg. 3
3. E-mail correspondence with David and Debi Barka, 3 February – 11 March 2013
4. Johnson, Caleb, *Alan B. Shepard, Jr.*, from genealogical website at http://www.mayflowerhistory.com
5. *Spartanburg Herald-Journal*, South Carolina, uncredited article "Alan Shepard was Happy-Go-Lucky Lad, Mother Says," issue Sunday, 24 June 1962, pg. 17
6. French, Francis and Colin Burgess, *Into That Silent Sea*, University of Nebraska Press, Lincoln, NE, 2003
7. *Spartanburg Herald-Journal*, South Carolina, uncredited article "Alan Shepard was Happy-Go-Lucky Lad, Mother Says," issue Sunday, 24 June 1962, pg. 17
8. *The Windsor Star* newspaper, (Ontario, Canada) article, Astronaut "Rugged Type," issue 5 May 1961, pg. 4
9. Carpenter, S., Cooper, Jr. L, Glenn, Jr., J., Grissom, V., Schirra, Jr., W., Shepard, Jr., A., and Slayton, D., *We Seven*, Simon and Schuster Inc., New York, NY, 1962, pg. 82
10. *The Pittsburgh Press* newspaper, article, "Astronaut Shepard Thrives on Danger," issue 5 May 1961, Pg. 8
11. Thomas, Shirley, *Men of Space, Vol. 3: Alan Shepard*, Chilton Company, Philadelphia and New York, 1961, pg. 189
12. Carpenter, S., Cooper, Jr. L, Glenn, Jr., J., Grissom, V., Schirra, Jr., W., Shepard, Jr., A., and Slayton, D., *We Seven*, Simon and Schuster Inc., New York, NY, 1962, pg. 83
13. *Spartanburg Herald-Journal*, South Carolina, uncredited article "Alan Shepard was Happy-Go-Lucky Lad, Mother Says," issue Sunday, 24 June 1962, pg. 17
14. Carpenter, S., Cooper, Jr. L, Glenn, Jr., J., Grissom, V., Schirra, Jr., W., Shepard, Jr., A., and Slayton, D., *We Seven*, Simon and Schuster Inc., New York, NY, 1962, pg. 83

15. Carpenter, S., Cooper, Jr. L, Glenn, Jr., J., Grissom, V., Schirra, Jr., W., Shepard, Jr., A., and Slayton, D., *We Seven*, Simon and Schuster Inc., New York, NY, 1962, pg. 85
16. Burgess, Colin, *Selecting the Mercury Seven: The Search for America's First Astronauts*, Springer-Praxis, Chichester, U.K., 2011
17. Chappell, Carl L., *Seven Minus One: The Story of Gus Grissom*, New Frontier Publication, Mitchell, IN, 1968, pg. 105
18. Shepard, Alan B., article, "First Step to the Moon," published in *American Heritage Magazine*, July/August 1994, issue Vol. 45, No. 4
19. Slayton, Donald K., *Pilot Training and Preflight Preparation*, extracted from National Aeronautics and Space Administration, National Institutes of Health and National Academy of Sciences report, *Results of the First U.S. Manned Suborbital Space Flight*, NASA HQ, Washington, D.C., 6 June 1961
20. Slayton, Donald K. and Michael Cassutt, *Deke: U.S. Manned Space from Mercury to the Shuttle*, Forge Books, New York, NY, 1994
21. Augerson, William S., M.D., *Physiological Responses of the Astronaut in the MR-3 Flight*, extracted from National Aeronautics and Space Administration, National Institutes of Health and National Academy of Sciences report, *Results of the First U.S. Manned Suborbital Space Flight*, NASA HQ, Washington, D.C., 6 June 1961
22. Schirra, Wally and Richard Billings, *Schirra's Space*, Naval Institute Press, Annapolis, MD, 1988, pg. 70
23. Carpenter, S., Cooper, Jr. L, Glenn, Jr., J., Grissom, V., Schirra, Jr., W., Shepard, Jr., A., and Slayton, D., *We Seven*, Simon and Schuster Inc., New York, NY, 1962, pg. 203
24. Interview with Dee O'Hara conducted by Colin Burgess and Francis French, San Diego, CA, 18 January 2003.
25. Charon, Mona, "Space Program Memories", *The Sun-Journal* newspaper, Lewiston, Maine, 3 August 1994, pg. 3
26. Wendt, Guenter and Russell Still, *The Unbroken Chain*, Apogee Books, Ontario, Canada, 2001, pg. 30
27. Schirra, Wally and Richard Billings, *Schirra's Space*, Naval Institute Press, Annapolis, MD, 1988, pg. 70
28. Cooper, Gordon and Bruce Henderson, Leap of Faith: An Astronaut's Journey into the Unknown, Harper Collins Publishers, New York, NY, 2000, pg. 21
29. Wolfe. Tom, *The Right Stuff*, Farrar, Straus and Giroux, New York, NY, 1979, pg. 215
30. Shepard, Alan, interviewed by Roy Neal for JSC Oral History program, Pebble Beach, Florida, 20 February 1998
31. French, Francis and Colin Burgess, *Into That Silent Sea*, University of Nebraska Press, Lincoln, NE, 2003
32. Shepard, Alan B., article, "First Step to the Moon," published in *American Heritage Magazine*, July/August 1994, issue Vol. 45, No. 4
33. *Ibid*
34. *Ibid*
35. Wendt, Guenter and Russell Still, *The Unbroken Chain*, Apogee Books, Ontario, Canada, 2001, pg. 32–33

4

Countdown to launch

At 12:30 a.m. on 2 May 1961 the countdown for MR-3 began, but the prospects of a launch were never very good. In spite of being late spring, violent thunderstorms had rumbled over the Cape that evening; it was raining heavily and occasional lightning flashes danced up and down the Florida coastline. Lining the beaches were people determined to see the launch, shivering under raincoats and ponchos, praying for the filthy weather to clear. The loading of liquid oxygen into the Redstone went ahead, but as the minutes ticked by the odds against flying steadily increased.

THUNDER OVER THE CAPE

Following the unexpected orbital mission of Yuri Gagarin and the nation's growing eagerness for an American to be launched into space, NASA had decided that as they were a civilian space agency and Mercury was an open program – unlike that of the Soviet Union – they would permit each flight to be televised live. On being assured about the abort system's capabilities, the thoroughness of the training, the readiness of the astronaut, and the integrity of the hardware involved, President Kennedy had agreed that the world should see the launch live. Nevertheless, he and his advisors remained concerned about a catastrophe in which the astronaut was lost as the world looked on. The television coverage was scheduled to commence at T-2 minutes and, around the nation, families settled in front of their television sets in a state of nervous excitement.

As arranged, flight surgeon Bill Douglas stole into crew quarters at 1:00 a.m. and gently roused the sleeping astronaut along with his backup, John Glenn. In response to Shepard's mumbled query, Douglas informed him the weather was quite bad at the moment, but a decision had been made to begin fueling the rocket. They were to proceed with the medical examination and suiting-up in the hope the weather would clear. After a shower and shave, Shepard donned his dressing gown and then joined Grissom and Glenn for an early morning high-protein breakfast of orange juice, filet mignon wrapped in bacon, and scrambled eggs, although the others opted instead for poached eggs.

104 Countdown to launch

Then, ready to face whatever lay in store for him that morning, it was time to undergo the pre-flight physical and psychiatric examinations.

Medical tests established that Shepard was still in the same perfect health he had previously enjoyed; his eyesight was normal, there were no respiratory ailments, his ear canals were clear, his thyroid was smooth and without any tenderness, his heart rhythm was regular and his blood pressure gave no discernible hint of the challenges he might soon face.

Then a lengthy interview with the NASA psychiatrist confirmed the astronaut's mental preparedness. The psychiatrist's report stated, in part, that Shepard "appeared relaxed and cheerful. He was alert and had abundant energy and enthusiasm. [His manner] was appropriate. He discussed potential hazards of the flight realistically and expressed slight apprehension concerning them. However, he dealt with such feelings by repetitive consideration of how each possible eventuality could be managed. Thinking was almost totally directed to the flight. No disturbances in thought or intellectual functions were observed." [1]

The next phase of the pre-launch operations for Shepard entailed having an array of medical sensors attached to his body. Clad only in the bottom half of his specially padded and ventilated long-johns, he stood patiently while the doctors positioned the

Conditions at the Cape on 2 May 1961 were never conducive for a launch. (Photo: NASA)

six sensors which would monitor and transmit his physiological state to the Mission Control Center. In this procedure, doctors glued a non-conducting cup containing a non-irritating paste to his skin, and used this paste as the lead-off from the skin. A shielded wire attached to a stainless steel mesh was buried within the paste, but not touching the skin. Four of these sensors went in predetermined spots under the right armpit, on the upper and lower chest, and on the lower left side of his body. Another was inserted into Shepard's rectum to record his anal temperature, and the last went below the nostrils to monitor his respiration. The six sensor wires were then bunched together in the common terminal that would later be plugged into a socket located adjacent to his right knee in the spacecraft.

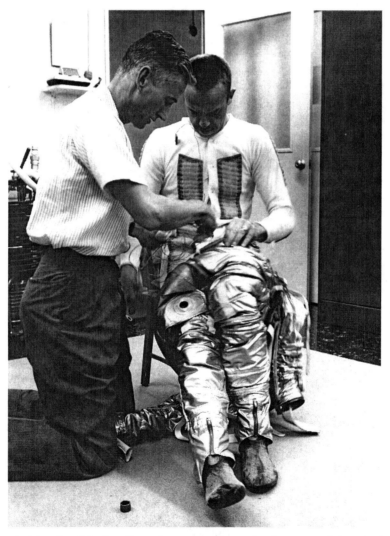

Suit technician Joe Schmitt adjusts the sensor wire socket incorporated into the leg of Shepard's suit. (Photo: NASA)

Next, Shepard made his way into the dressing room where NASA suit technician Joe Schmitt was waiting to assist in the awkward procedure of donning his $10,000 pressure suit and helmet. The first item of clothing was long cotton underwear with ribbed sections on the arms, legs, and back to facilitate the circulation of air. Next came the custom-made, 20-pound silvery space suit itself, which was comprised of an inner layer of rubber and an outer layer of aluminized nylon. The suit was sealed by means of airtight zippers, laces, and straps, and encircling the neck area was a soft rubber cone that would make the suit waterproof when the helmet had been removed for egress. Although the suit was inflatable, this safeguarding measure would only be taken if there was a loss of air pressure inside the capsule, and it would allow the astronaut 90 valuable minutes of protection. Once inflated, the suit became almost rigid, although the gloves were designed with curved fingers to allow the astronaut to grip the controls, albeit with the exception of a single finger on the left glove, which remained straight for the purpose of pushing buttons.

Shepard took a seat, and after both legs of the suit had been inserted one at a time, the bundled sensor wires were carefully threaded through a hole in the thigh area. He then stood up and slipped his arms one at a time into the sleeves of the suit, which was zippered across his chest and middle. After pulling boots over his white socks and securing them, Shepard pulled on his gloves and zipped them to each sleeve. He also slipped on a pair of plastic overshoes which he would remove prior to insertion into the spacecraft. These were to prevent dirt picked up on the way to the hatch of *Freedom 7* from entering the capsule. Finally, Joe Schmitt lowered Shepard's helmet over his head, securing it in place with a ring lock.

An air tightness test was then conducted on the space suit. To accomplish this, Shepard reclined in a contour couch and closed his helmet's faceplate. Schmitt then inflated the suit to 5 psi and checked it for leaks. Finding none, the suit was deflated again and a portable air-conditioning unit was connected to the suit. Shepard would carry this in his hand from that moment until he was ready to plug himself into the spacecraft's air-conditioning system.

The preparations complete, Shepard then sat in Hangar S awaiting the go/no-go decision, his suit cooled by the air-conditioning unit.

"The signs were not propitious," he later explained. "And at 3:30 a.m., with the liquid oxygen already loaded aboard the booster, the technicians took a look at the lightning and declared a 'hold.' They started working again at 3:50, with the count at T minus 290 minutes." [2]

DIMINISHING CHANCES

Despite the ominous weather, flight preparation work continued at Pad 5. As backup pilot, John Glenn realized that there was no immediate prospect of replacing a fit and ready Shepard on the flight, even if the storm abated. He eluded the waiting press and headed off to the pad to assist in preparing *Freedom 7*.

Meanwhile, Shepard, remaining in Hangar S, was informed that two ten-minute pauses had occurred in the lengthy countdown in order to assess weather reports. On the nominal schedule, he was to make the three-mile-long journey by transfer van to the pad at 4:00 a.m., but that time came and went, and soon he was watching the first pink tendrils of dawn tinting the gray clouds in the eastern sky. With no word from the weather people, Shepard was coming to the realization that the squall line, which lay ahead of a cold front stretching from Virginia to the Gulf coast, would probably prevent a launch that day.

"I frankly didn't think we would go that morning. I wasn't trying to second-guess anyone, but the weather did not look good at all. I was sure we wouldn't get the results we needed, even if we did go. But the crews were ahead on the countdown, and if we didn't try that morning we would have to go through a long 48-hour delay before we could refuel the Redstone and try again." [3]

A fully suited Alan Shepard bides his time, waiting to fly. (Photo: NASA)

Outside, a small group of authorized reporters and photographers representing the vast media army gathered at the Cape were also checking their watches, anxious for something to happen. For some, their main objectives were to photograph or film the space-suited astronaut leaving the hangar for the transfer bus, and the reporters were eager to communicate every move back to their editors. What everyone wanted to know, was which one of the three nominated astronauts was going to fly? NASA had still not announced whether it would be Glenn, Grissom, or Shepard, but the betting was on the affable Marine, John Glenn.

Then, suddenly, it was all over for that day; two storm fronts were converging on the Cape and down along the 290 miles of the Atlantic Missile Range over which the Redstone would fly. As the decision came, Shepard was standing just inside the door

The public could only speculate on which of the three main candidates would fly the MR-3 mission. (Photo: NASA)

of the hangar, seconds away from going out to the transfer van. He was disheartened by the news, but not surprised. The launch had been postponed for at least two days. NASA needed clear visibility for the mission, especially in the critical first minutes, because the flight controllers would require good visual tracking in order to be ready to trigger Shepard's escape mechanism at the first hint of trouble. That, they decided, was not going to be the case.

At 7:40 a.m., just 2 hours 20 minutes before the planned liftoff, an announcement came over loudspeakers that the shot had been postponed. "No new launch date has been set, but the minimum recycle time is 48 hours. The pilot will remain in the crew quarters in the Mercury hangar here."

NASA's Chief of Public Information, John ('Jack') King officially informs media representatives that the MR-3 flight has been scrubbed. (Photo: Associated Press)

After Shepard had doffed his spacesuit, he was given a small glass of brandy to help him over his disappointment. "He didn't really need it," according to Lt. Col. John ('Shorty') Powers, NASA's Public Affairs Officer. "There were about nine of us there who needed it more than he did. He just joined us." [4]

POSTPONEMENT

It was a frustrating time for the reporters and photographers, and for the public now deserting the Cape's sodden beaches. They had all spent a wet and miserable night waiting for the eagerly anticipated launch shortly after sunrise.

But in the midst of the bad news, there was an unexpected revelation: NASA had decided to reveal the name of the first astronaut. The announcement stated that Cdr. Alan Shepard had been selected to pilot the flight that day, and this would probably remain so for the next attempt. "I was relieved when they made the announcement," Shepard later revealed. "It was getting to be a strain keeping the secret." [5] Ironically, just thirty minutes after the delay announcement, the Sun broke through the dense cloud layer.

Like everyone else, the news media could only watch and wait. (Photo: NASA)

Apart from some maintenance work on the vehicle, everything remained in a 'go' situation. However, the cold front that had stationed itself over the Florida peninsula continued to keep launch conditions below the required minimum. Over the next two days, technicians painstakingly purged the Redstone of its corrosive fuel, rechecked its circuitry, and carried out a repair to one of the liquid oxygen lines.

Meanwhile, apart from some simulator work, Shepard was able to relax; taking a nap, answering mail, running at a local beach, and going over the flight plan with his backup and roommate, John Glenn. The weather slowly began to improve, leading Col. Powers to inform a bevy of anxious reporters, "The weather man tells us that it looks like the weather will be clear enough for us to go ... the chances are better than 50-50 in our book that we can get off the launching before the weather worsens." [6]

Shepard was a relieved man. "At the scheduled meeting Thursday morning we got pretty fair weather reports. The launch crews were picking up the count again at T minus 390 minutes, and I felt glad that I was going to be able to give it a whirl." [7]

Pad 5 as seen from the blockhouse on 29 April during an emergency egress exercise. In a pad abort, Shepard would escape by operating the mechanically actuated side hatch, discarding it, and then scrambling into the basket of the articulated "cherry-picker" crane. (Photo: NASA)

The three-day delay actually proved beneficial to the waiting astronaut. "To my surprise, I felt the launch delay actually eased the tension that had been building up inside me. Before the May 2 [attempt] I'd been plagued with visions of rockets tumbling out of control or blowing up in the air – after all, I'd seen this happen – but during those three days I was able to back off, regroup, and hit it again. I recognized I was experiencing normal apprehension and not fear. The entire reasoning process was old hat to a test pilot. I knew how to turn off this kind of stuff, and I felt calm as the new launch date of May 5 neared." [8]

A pre-flight briefing was conducted at 11:00 a.m. on 4 May in order to examine all the operational elements of the flight. "This briefing was helpful since it gave us a chance to look at weather, radar, camera, and recovery force status. We also had the opportunity to review the control procedures to be used during flight emergencies as well as any late inputs of an operational nature. This briefing was extremely valuable to me in correlating all of the details at the last minute." [9]

That afternoon Shepard and Glenn took a leisurely walk along a nearby beach to catch crabs. They were ready to go.

"The night of May 4, however, the other astronauts and support teams brought their own tension onto the scene," Shepard reflected. "Everyone but me was walking on eggshells. Despite the strong feelings about weather, rocket reliability, the escape system, anything and everything, no one dared broach those subjects. It all got so thick that I went into my bedroom and phoned my family in Virginia Beach." [10]

Louise was delighted to hear from her husband. They discussed the weather and the prospects for a launch the next morning. He spoke briefly with Louise's parents and his daughters before promising his wife he would take care of himself and that he loved her. Then he went to get some sleep.

A DAY FOR HISTORY

It was 1:10 a.m. when flight surgeon Bill Douglas gently woke Shepard. "Come on, Al," he said. "They're filling the tanks."

Shepard had only been asleep for three hours, but was instantly awake and alert. "I'm ready," he replied. "Is John awake?" Douglas saw that John Glenn was already clambering out of his bed, ready for whatever the day would bring.

"John's awake," Douglas confirmed. "We're all awake. Did you sleep well?"

Shepard said that he had slept soundly and had no recollection of any dreams. He added that upon awakening around midnight he had peeped out through the window to see if it was still raining. "The stars were out and I went back to sleep," he pointed out with a slight smile. The weather, he was told, was indeed looking good.

Whistling quietly to himself, Shepard walked to the bathroom where he shaved and showered, then in company with Dr. Douglas and Glenn polished off a breakfast of filet mignon, eggs, orange juice, and tea. "I left the breakfast table to place myself at the mercy of the doctors," he subsequently recorded, "who did their usual poking, prodding, and measuring, and then attached a battery of sensors to me." [11] While Douglas and Grissom remained with Shepard, Glenn dressed and went to the launch pad to once again check out the spacecraft.

A day for history   113

Dietician Capt. Jean McKay serves a launch-day breakfast to Shepard and Glenn. (Photo: NASA)

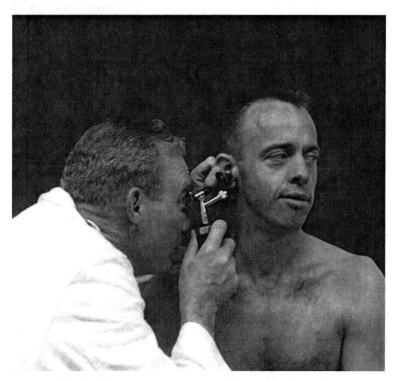
Astronaut physician Lt. Col. Dr. William Douglas inspects Shepard's ears. (Photo: NASA)

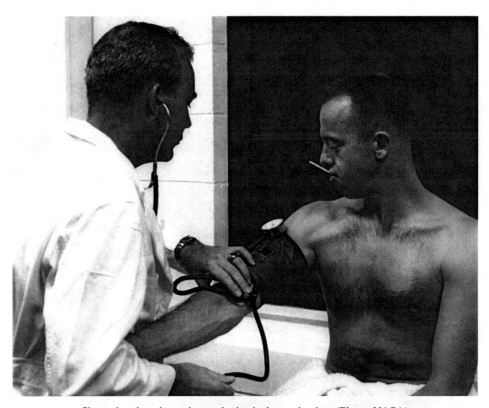

Shepard undergoing a thorough physical examination. (Photo: NASA)

Altogether, the medical and psychiatric assessments took a little under two hours, but they showed that Shepard was in excellent physical and mental shape for the flight. He had a slight sunburn on his shoulders from staying out in the Sun too long at a swimming pool, and a blackened toenail from Grissom accidentally stepping on his foot a few days before. Most importantly, his respiration and blood pressure were good, while his pulse rate was measured at 75 beats per minute.

The psychiatric examination lasted around an hour, at the end of which the psychiatrist reported, "He realized the dangers he was about to face, but showed no fear. Never seen a man so calm. I tried to get him to talk about other things than the flight, about his family, for example, to see whether this would make him anxious, but I didn't succeed. All his mind, every nerve, was concentrated on the flight: nothing else interested him. Even while on his way to the suit room he was already a part of the spacecraft." [12]

Shepard was then assisted in donning his space suit. Dee O'Hara had already been busy that morning assisting Bill Douglas with his medical checks and procedures, and she vividly recalls the suiting-up time for Shepard. "I remember when he walked into the suit room to get suited up, it was just … everything became dead silent, just became very quiet, and Deke was there, and everybody just sort of milled around, and not much was said. There was hardly any conversation. Joe Schmitt did his checks, and

A day for history 115

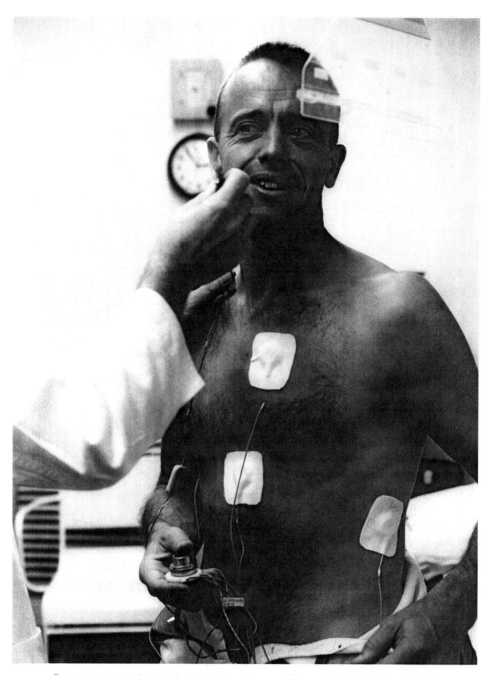

Sensors were taped to predetermined positions on Shepard's body. (Photo: NASA)

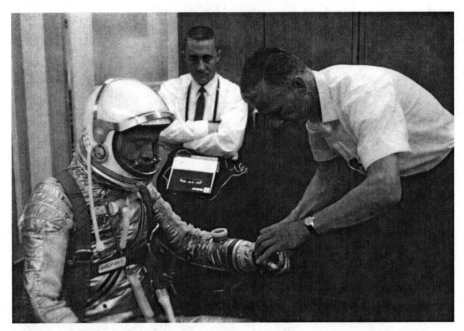
Suit technician Joe Schmitt makes final adjustments to Shepard's gloves. (Photo: NASA)

Alan was getting into his boots and, you know, whatever … but it was just dead quiet that day." [13]

Observing this lengthy process, Bill Douglas was mentally drawing comparisons with an earlier event. "I don't quite know why," he reflected, "but it reminded me of the dressing of the matador before the *corrida*. An astronaut and a matador have nothing in common, but once I was in Spain and I was present at the dressing of a matador and the atmosphere was the same: a solemn anxiety, a religious silence, a lot of people around him. And over everything a vague smell of death." [14]

Once Shepard was clad in his suit and helmet, a pressure check was carried out by technician Joe Schmitt. Once its integrity had been verified, the suit was deflated; it would not be reinflated until just prior to launch.

At 3:55 a.m., carrying his portable air-conditioning unit, Shepard began to make his way downstairs. Footage from the ABC network television coverage shows Dee O'Hara in a window above the exit. She accompanied him as far as the hall, where he turned to her and said, "Well, Dee, here I go." Then he followed Joe Schmitt out through the hangar door.

"I was very, very frightened," O'Hara revealed to the author. "Particularly when he left and went downstairs to get to get in the van. I didn't know if I was going to see him again, and I just … I straightened up the area." [15]

Immediately that the hangar door was opened, flashbulbs began to pop and TV cameras followed the astronaut in his silvery suit as he walked behind Schmitt to the small transfer van and cautiously stepped in. They were joined by Bill Douglas, Gus Grissom, and several technicians. Joe Schmitt was there to assist Shepard into his restraint harness once he had been inserted into the capsule.

A day for history 117

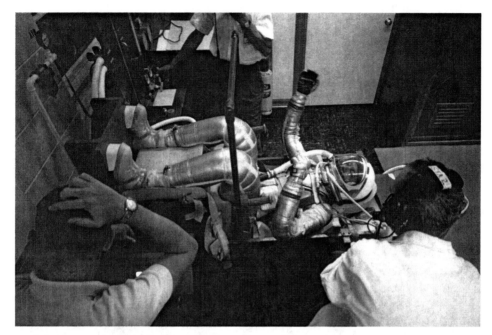

Suit technician Joe Schmitt checks the pressurization of Shepard's space suit, with Dr. Douglas (with Station 2 headset) observing the procedure. (Photo: NASA)

Forty minutes later, the van pulled up alongside the launch gantry, essentially a modified oil derrick. It was the same launch pad from where America's first satellite, *Explorer 1*, was launched into orbit some three years earlier. There was still some time to kill before he could enter the capsule. Gordon Cooper entered the van to give Shepard a final update on the weather and on the positions of the recovery ships. "He said the weathermen were predicting three-foot waves and 8-to-10-knot winds in the landing area, which was within our limits," Shepard later recorded. "We had a device in the van to check on the sensors, and everything was working fine. I rested my weight in a reclining chair while all this was going on." [16]

In order to ease some of the nervous anticipation felt by all in the van, Al and Gus spontaneously broke into their favorite Bill Dana routine, with Shepard playing the role of the reluctant astronaut, José Jiménez. Part of the well-known routine involved José listing all of the qualities an astronaut ought to have, such as courage, perfect vision, and low blood pressure. Then he finished with, "And you got to have four legs." Grissom, playing the straight man, asked, "Why four legs?" Shepard grinned widely, and in his best José imitation responded, "They really wanted to send a dog, but they thought that would be too cruel!" It did the job, and Shepard was in a good mood when he was informed that it was time to leave the van [17].

The door was opened, and Shepard carefully climbed down four steps onto solid concrete. Above him the sky was still dark, with a thin sliver of Moon peeping out from small dark clouds. Bright searchlight beams cut back and forth, while arc lights vividly lit the area. But he only had eyes for one thing that morning. He took in the

The moment everyone had been anticipating, as Alan Shepard departs Hangar S for the launch pad. (Photo: NASA)

A day for history 119

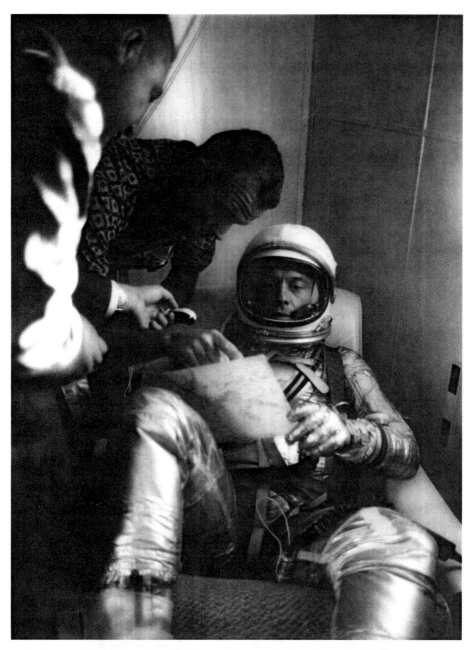

As Grissom (left) looks on inside the transfer van, Gordon Cooper briefs Shepard on the prevailing weather conditions. (Photo: NASA)

gleaming Redstone emblazoned in the brilliant glare of the searchlights, rimmed with frost and ice and gently issuing swirling vapors of liquid oxygen. Around the foot of the rocket, moving through the clouds of vapor, dozens of engineers and technicians were engaged in final preparations, wearing construction hard hats of various colors to denote their work.

Shepard steps down from the transfer van. (Photo: NASA)

Gazing up at the Redstone, Shepard pauses on his way to the gantry elevator. (Photo: NASA)

"I stepped out into a strange world of glaring floodlights and banshee wails from a breeze blowing across supercold fuel lines," Shepard would later recall. "I looked up, for the moment overwhelmed by the gleaming blue-white lights. Then I began the final walk toward the gantry elevator. 'Up' was six stories above me." [18]

Then Shepard paused at the gantry base, along with Grissom and Dr. Douglas. He shaded his eyes with his left hand and looked up, taking in the sight of the rocket that he would soon ride into the heavens.

"I sort of wanted to kick the tires – the way you do with a new car or an airplane. I realized that I would probably never see that missile again. I really enjoy looking at a bird that is ready to go. It's a lovely sight. The Redstone with the Mercury capsule and escape tower on top of it is a particularly good-looking combination, long and slender. And this one had a decided air of expectancy about it. It stood there full of LOX, venting white clouds and rolling frost down the side. In the glow of the searchlight it was really beautiful." [19]

After boarding the elevator at 5:15 a.m., Shepard turned and waved at the launch team, who were cheering loudly and applauding. He had meant to stop and express his thanks, but the emotion of the moment got to him. As they ascended the 70 feet to the level known as "Surfside 5" where he would ingress *Freedom 7*, Bill Douglas unexpectedly handed Shepard a small gift from a good friend, NASA engineer Sam Beddingfield. It was a box of crayons. They'd once shared a joke about an astronaut about to start a long mission who had taken along a coloring book to help him pass the time, but refused to fly when he found that he had forgotten his Crayolas. "Just so you'll have something to do up there," Douglas said with a wide smile. Shepard laughed as he handed the crayons back, saying he might just be a little too busy to use them.

They exited the elevator and made their way into the green-colored gantry room (curiously known as the "White Room") where the spacecraft stood ready for him to climb in through the two-foot square hatchway and prepare to make history. Shepard

Shepard, Douglas, and Grissom board the gantry elevator (Photo: NASA)

walked around a little, talking briefly with Glenn and Grissom, thanking them for all their hard work, especially Glenn – now wearing a pristine white coat and cap – who had served as his backup pilot. As he moved over to the hatch, he looked once again at the unadorned name boldly painted on the side of his spacecraft. "My choice," he would explain. "*Freedom* because it was patriotic. *Seven* because it was the seventh Mercury capsule produced. It also represented the seven Mercury astronauts." [20]

At 5:18 a.m., after Glenn had made a final visual check of the spacecraft interior, Shepard gripped his hand in a hearty handshake, and then began the delicate task of inserting himself into the cramped confines of the capsule. McDonnell engineers first assisted the astronaut in removing his protective overshoes, then he lowered the visor of his helmet and wormed feet-first in through the hatch.

"My new boots were so slippery on the bottom that my right foot slipped off the right elbow of the couch support and on down into the torso section, causing some superficial damage to the sponge rubber insert – nothing of any great consequence, however. From this point on, insertion proceeded as we had practiced. I was able to get my right leg up over the couch calf support and part way across prior to actually getting the upper torso in. The left leg went in with very little difficulty … I think I had a little trouble getting my left arm in, and I'm not quite sure why. I think it's mainly because I tried to wait too long before putting my left arm in. Outside of that, getting into the capsule and the couch went just about on schedule, and we picked up the count

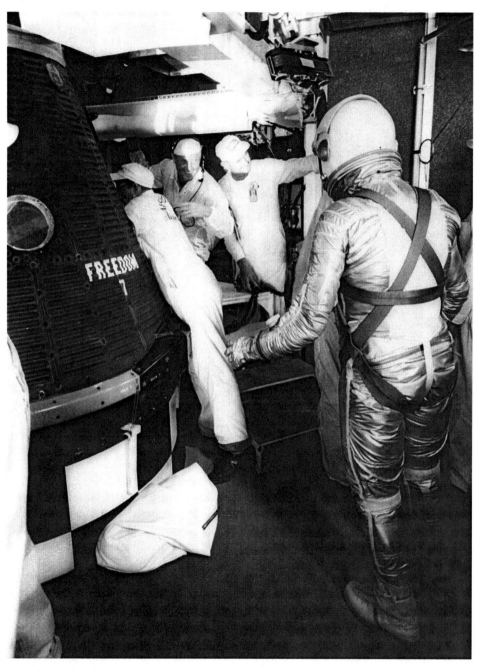

A grinning John Glenn welcomes Shepard to the White Room. (Photo: NASA)

Shepard offers a last thank-you to Gus Grissom. (Photo: NASA)

for the hooking up of the face plate seal, for the hooking up of the biomed connector, communications, and placing of the lip mic[rophone]. Everything went normally." [21]

Joe Schmitt had one final role to play during Shepard's insertion into *Freedom 7*, and it all went as planned. "I had been training with him for so long. I mean that's all we had been doing …. My job was not only to suit them and take care of the suits, but also to put them in the spacecraft and hook up their communications, their hoses, and also their restraint straps." [22]

Part of the ingress procedure required Schmitt to first remove an instrument panel, allowing Shepard enough room to slide in and nestle into his contour couch before Schmitt replaced the panel and attended to the restraint straps and his other pre-flight tasks.

After being physically connected with his capsule, Shepard noticed a stray slip of paper amongst his instruments which read, in the handwriting of John Glenn, "Ball games forbidden in this area." He laughed at this little bit of levity and handed it out to the smiling Marine, who then set off to the Mercury Control Center.

Already strapped in position on a small ledge inside the spacecraft was something Shepard hoped he would not have to use – a parachute chest pack. It was there in the event of a serious problem with the main parachute prior to landing. If necessary, he

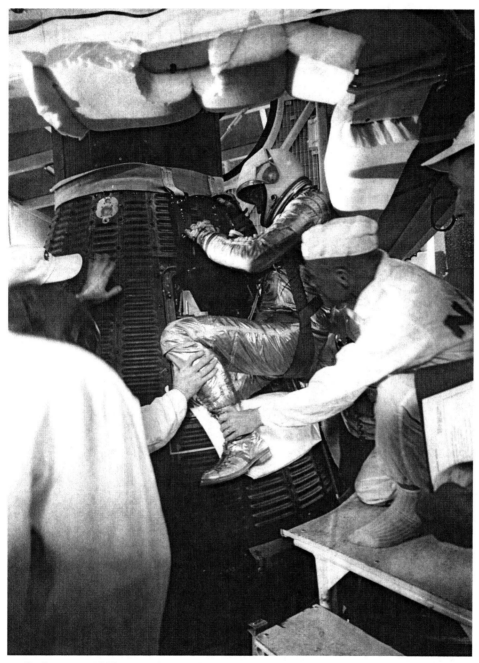

Having removed his protective overshoes, Shepard is eased into *Freedom 7* with the assistance of backup pilot John Glenn. (Photo: NASA)

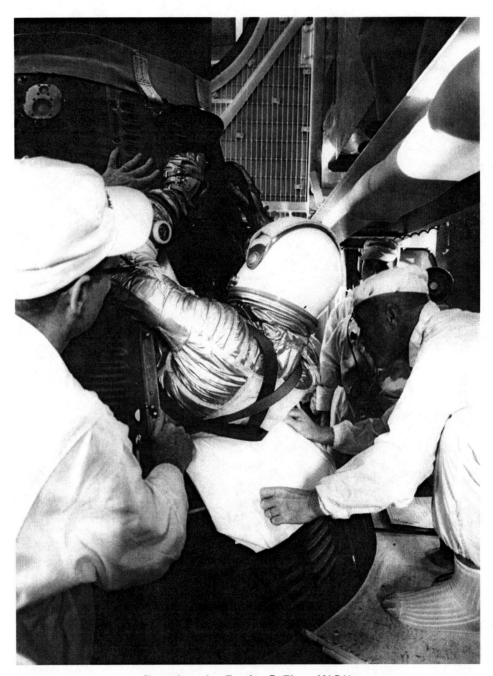

Shepard entering *Freedom 7*. (Photo: NASA)

was to clip it on, manually operate and discard the mechanically actuated side hatch, then squeeze out of the rapidly descending capsule. But Shepard knew it would be an extremely difficult task extracting himself from the couch, opening the hatch, and scrambling out in time, so even though he took note of it in his checks, he quickly dismissed its presence and purpose from his mind. Although the bulk of the personal parachute made the interior of the capsule even more cramped than necessary, the planners nevertheless loaded it on this flight and the subsequent MR-4 mission.

"The preparations of the capsule and its interior were indeed excellent," Shepard would observe in the MR-3 post-launch report. "Switch positions were completely in keeping with the gantry check lists. The gantry crew had prepared the suit circuit purge properly. Everything was ready to go when I arrived, so, as will be noted elsewhere, there was no time lost in the insertion. Insertion was started as before.

"After suit purge, the suit-pressure check showed no gross leaks; the suit circuit was determined to be intact, and we proceeded with the final inspection of the capsule interior and the removal of the safety pins. I must admit that it was indeed a moving moment to have the individuals with whom I've been working so closely shake my hand and wish me *bon voyage* at this time."

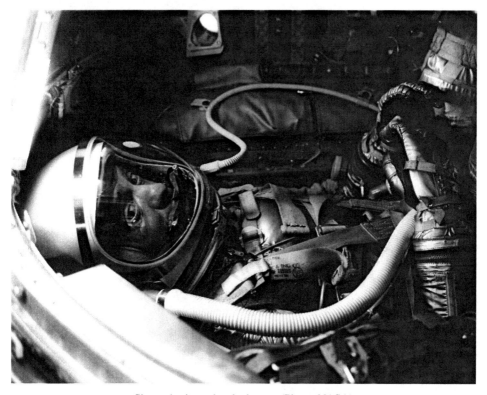

Shepard prior to hatch closure. (Photo: NASA)

A final glimpse of the astronaut as the capsule's hatch is closed. (Photo: NASA)

At 6:10 a.m., the pad technicians began the task of installing the spacecraft hatch, which was held in place by 70 bolts. The ensuing cabin leak check was completed to everyone's satisfaction. Shepard's training now kicked in as he began industriously working through his checklists, ensuring once again that everything was exactly as it should be, and that all the switches were at the correct settings.

Shepard later reported, "The point at which the hatch itself was actually put on seemed to cause no concern, but it seemed to me that my metabolic rate increased slightly here. Of course, I didn't know the quantitative analysis, but it appeared as though my heart beat quickened just a little bit as the hatch went on. I noticed that my heart beat, or pulse rate, came back to normal again shortly thereafter with the execution of normal sequences. The installation of the hatch, the cabin purge, all proceeded very well, I thought. As a matter of fact, there were very few points in the capsule count that caused me any concern." [23]

Every so often Shepard glanced into the periscope to monitor the outside activity, and would smile to himself whenever the wide-angle-lens-distorted grinning face of Grissom filled the small screen. As the White Room crew went about their business, little did Shepard realize that he would spend the next four hours on his back in the form-fitting couch, as delay after delay threatened the increasingly irritated astronaut with yet another launch scrub.

A day for history 129

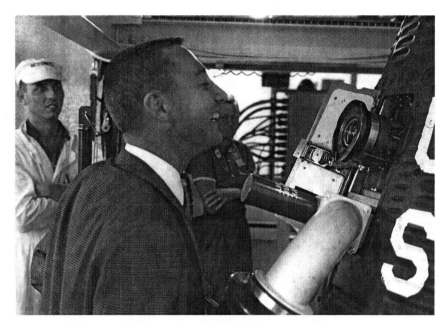

A cheeky Grissom peers into the capsule's periscope. (Photo: NASA)

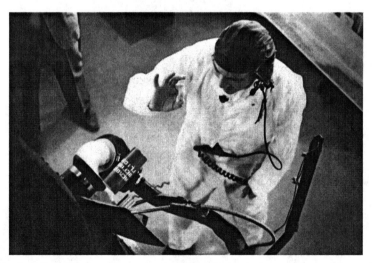

Physician Bill Douglas gives Shepard a final "okay" sign in front of the periscope. (Photo: NASA)

Then the voice of CapCom Deke Slayton came through. "José? Do you read me, José?"

"I read you loud and clear, Deke," Shepard replied.

"Don't cry too much," Deke said as part of the Bill Dana routine.

"All right," came the more sober response.

At 6:34 a.m., some 24 minutes after Shepard had been sealed into *Freedom 7*, the enclosing service structure slowly began to roll away from the Redstone, leaving the impressive white-and-black painted rocket poised pencil-like on the launch pedestal, pointed ambitiously towards the rapidly lightening dawn sky, ready to lunge free on command.

There was now an air of hope and expectation among the delay-weary hordes of reporters and members of the public who were again on the beaches and every other vantage point, listening to bulletins on their transistor radios. They had endured three frustrating and exhausting days of storms sweeping up and down the coast, thunder and lightning, and the dispiriting announcements of continued postponements. Now, as they assembled beneath a relatively cloudless dawn sky, they began to believe that this might, finally, be the day on which Alan Shepard would make history.

DELAY AFTER DELAY

As dawn broke over the New Hampshire hills, the first pale rays of the Sun fell on a crisp new American flag that had been proudly raised earlier that morning by Renza Shepard on the front lawn of their home in East Derry.

The launch gantry begins to roll away. (Photo: NASA)

Aboard the aircraft carrier USS *Lake Champlain* (CVS-39), the prime recovery ship, the crew stood silently in the dawn light as Rev. Henry Faville Maxwell, their chaplain, intoned a heartfelt prayer for the success of Shepard's mission over the ship's loudspeakers.

"Dear Lord who hears us, now that a precious life is about to be flung into the heavens, we are filled with fear; we are afraid of imminent danger. Dear Lord who hears us, we thank Thee for giving us men ready to sacrifice their existence to open up for us the doors of space. May he succeed without losing his life. May success crown his endeavors to explore the paths of knowledge; not only that we may expand into the universe, but that it will be a peaceful universe where we live with each other and with Thee. Amen." [24]

In the Pad 5 blockhouse, fellow astronaut Gordon Cooper communicated with the capsule until this task transferred to Deke Slayton in the Mercury Control Center. By then, Slayton had been joined by Flight Director Chris Kraft and Operations Director Walt Williams.

As Williams later recalled of that day, "You can say that intuition means nothing, but there are days when you feel things are right and days when you feel things are wrong. On the previous Tuesday, the weather was a problem – but that was only one factor. We were having small problems – no serious ones, but things weren't going well. That is why I scrubbed quite early.

"On Friday, even though we had problems, I felt that we could handle each one as it came up. I was in the Mercury Control Center – at the back of the room. Before that, I had been roaming around to the pad, the blockhouse, up the gantry, in to see how Al was coming along. Once I put on my 'Operations Director hat,' I am in total charge. I don't mean that in an autocratic way, but someone has to call the shots. In essence, I answer to no one except the President. Once we are under way with the countdown, it is a minute-by-minute decision whether we go or not." [25]

Once the gantry had rolled on its tracks clear of the Redstone, Shepard began to feel more confident by the minute. "The periscope gave me a view of clouds lit by the morning Sun. Far below, I watched the launch crew finishing last-minute details at the base of the rocket. I glanced at my capsule timer. Only fifteen minutes to go. The view outside dimmed. Cloud cover rolling in. Damn!"

Now the brightness of the Sun was intruding into the cabin through the periscope, so Shepard cranked a couple of filters over the screen to diminish the glare.

The countdown clock stopped during the delay, and everyone began scanning the skies, eager for the clouds to depart. "Everyone hated countdown delays," Shepard later observed. "They just allowed more time for something to go wrong." [26]

And something did go wrong. An inverter, a small electrical part in the rocket that changed DC current to AC, developed a fault. It may have been a relatively minor thing, but it had to be fixed. Everything had to be fully operational for the mission to proceed. To everyone's disappointment, the launch director ordered the gantry rolled back in.

As Shepard observed later, "There was a time during the countdown when there was a problem with the inverter in the Redstone. Gordon Cooper was the voice communicator in the blockhouse. So he called and said, 'This inverter is not working in the Redstone. They're going to pull the gantry back in, and we're going to change inverters. It's probably going to take about an hour, an hour-and-a-half.' And I said, 'Well, if that's the case then I would like to get out and relieve myself.'

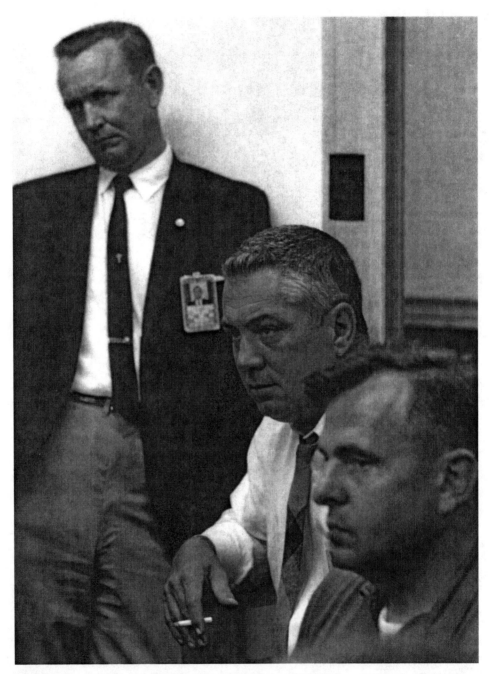

Flight Director Chris Kraft (at top), Operations Director Walt Williams, and Project Engineer Walter Kapryan at a pre-launch conference. (Photo: NASA)

Delay after delay 133

Mercury Control Center at the Cape prepares for the MR-3 launch. (Photo: NASA)

"We had been working with a device to collect urine during the flight that worked pretty well in zero-gravity but really didn't work very well when you were lying on your back with your feet up in the air, like you were on the Redstone. And I thought my bladder was getting a little full and, if I had some time, I'd like to relieve myself. So I said …

"Gordo?"
"Go, Alan?"
"Man, I got to pee!"
"You what?"
"You heard me. I've got to pee. I've been in here forever. The gantry is still right here, so why don't you guys let me out of here for a quick stretch?"

"Hold on," came Cooper's response. He consulted with Wernher von Braun and a few minutes later came back. "No way, Alan. Wernher says we don't have the time to reassemble the White Room. He says you're in there to stay."

"Gordo, I could be in here a couple more hours, and by that time my bladder's gonna burst!"

"Wernher says no."

"Well, shit, Gordo, we've got to do something. Dammit, tell 'em I'm going to let it go in my suit!"

"No! No, good God, you can't do that," Cooper called back. "The medics say you'll short-circuit all their medical leads."

"Tell 'em to turn the power off!"

The solution was that simple. "Gordo had a chuckle in his voice when he told me, 'Okay, Alan. Power's off. Go to it.' It was as if they'd designed the suit for such an emergency. In that semi-supine position the liquid pooled in the small of my back and my heavy undergarment soaked it up. With 100 percent oxygen flowing through the suit, I was soon dry. The countdown resumed. The gantry was gone." [27]

Knowing that his family was watching a live television transmission from the Cape, Shepard called Cooper in the blockhouse once again and requested that he get Shorty Powers to ring Louise and let her know he was fine despite the delay, which had extended to 1 hour 26 minutes and pushed back the projected time of liftoff.

Gordon Cooper communicating with Alan Shepard with Wernher von Braun looking on. (Photo: NASA)

An aerial view of the Pad 5 blockhouse at Cape Canaveral. (Photo: NASA)

Then, with 2 minutes 40 seconds remaining on the clock, technicians noticed that the fuel pressure in the Redstone was running a little high, and Shepard was warned there might be another short delay. Having heard enough of what he felt was a severe case of over-caution, there was a brittle snap to his voice when he responded. "Shit! I've been here more than three hours. I'm a hell of a lot cooler than you guys are. Why don't you just fix your little problem and light this candle!" [28]

Without being unkind towards Alan Shepard, and the way in which this incident is portrayed in the movie *The Right Stuff*, his words didn't galvanize the firing team into action – in fact, the blockhouse technician most involved, Andy Pickett, was not even in the capsule-blockhouse voice loop. He had noticed and reported an irregular reading on the propulsion regulator, which indicated a slight pressure increase. He then flicked a switch a couple of times to open and close a vent valve. This rectified the problem, and the pressure returned to normal within one minute of the anomaly being noticed. The countdown was resumed.

On hearing that the technical glitch had been fixed, Shepard gave a sigh of relief, then called Slayton, who had taken over direct communications from Gordon Cooper in the blockhouse.

"Are we ready, Deke?" he asked, and got the answer he wanted.

"Ready, Al."

The articulated "cherry picker" in place, ready to conduct an astronaut evacuation in the event of a launch mishap. (Photo: NASA)

The spindly "cherry picker" swung to its standby position by the Redstone, ready to move in and retrieve the astronaut in the event of a looming disaster.

The rocket was ready, the spacecraft was ready, the range was ready, and Shepard was most definitely ready; it was time to light the candle.

References

1. Jackson, Carmault B., Jr., M.D., William K. Douglas, M.D., James F. Culver, M.D., George Ruff, M.D., Edward C. Knoblock, Ph.D., Ashton Graybiel, M.D., *Results of Preflight and Postflight Medical Examinations (Part 5)*, from *Results of the First U.S. Manned Suborbital Space Flight*, National Aeronautics and Space Administration, the National Institutes of Health and National Academy of Sciences, published by NASA HQ, Washington, D.C., 6 June 1961
2. Potter, Sean, *Weatherwise* magazine article, "Retrospect: May 5, 1961: The First American in Space," Heldref Publications, Philadelphia, PA, issue May/June 2011
3. *Ibid*
4. Carpenter, S., Cooper, Jr. L, Glenn, Jr., J., Grissom, V., Schirra, Jr., W., Shepard, Jr., A., and Slayton, D., *We Seven*, Simon and Schuster Inc., New York, NY, 1962, pg. 238
5. *Ibid*, pg. 238
6. Potter, Sean, *Weatherwise* magazine article, "Retrospect: May 5, 1961: The First American in Space," Heldref Publications, Philadelphia, PA, issue May/June 2011
7. *Ibid*
8. Shepard, Alan B., article, "First Step to the Moon," published in *American Heritage Magazine*, July/August 1994, issue Vol. 45, No. 4
9. National Aeronautics and Space Administration, National Institutes of Health and National Academy of Sciences, *Results of the First U.S. Manned Suborbital Space Flight*, NASA HQ, Washington, D.C., 6 June 1961
10. Shepard, Alan B., article, "First Step to the Moon," published in *American Heritage Magazine*, July/August 1994, issue Vol. 45, No. 4
11. *Ibid*
12. Fallaci, Oriana, *If the Sun Dies*, Collins, London, U.K., 1967
13. Interview with Dee O'Hara conducted by Colin Burgess and Francis French, San Diego, CA, 18 January 2003.
14. Fallaci, Oriana, *If the Sun Dies*, Collins, London, U.K., 1967
15. Carpenter, S., Cooper, Jr. L, Glenn, Jr., J., Grissom, V., Schirra, Jr., W., Shepard, Jr., A., and Slayton, D., *We Seven*, Simon and Schuster Inc., New York, NY, 1962, pg. 240
16. *Ibid*, pg. 240
17. Shepard, Alan B., article, "First Step to the Moon," published in *American Heritage Magazine*, July/August 1994, issue Vol. 45, No. 4
18. Carpenter, S., Cooper, Jr. L, Glenn, Jr., J., Grissom, V., Schirra, Jr., W., Shepard, Jr., A., and Slayton, D., *We Seven*, Simon and Schuster Inc., New York, NY, 1962, pg. 241
19. Shepard, Alan B., article, "First Step to the Moon," published in *American Heritage Magazine*, July/August 1994, issue Vol. 45, No. 4
20. NASA Project Mercury Working Paper No. 192, *Post-launch Report for Mercury-Redstone No. 3 (MR-3)*, NASA Space Task Group, Langley Field, VA, 16 June 1961
21. *Ibid*

22. Schmitt, Joe, interview with Michelle Buchanan and Steven Spicer for JSC Oral History program, Friendswood, Texas, July 1997
23. NASA Project Mercury Working Paper No. 192, *Post-launch Report for Mercury-Redstone No. 3 (MR-3)*, NASA Space Task Group, Langley Field, VA, 16 June 1961
24. Fallaci, Oriana, *If the Sun Dies*, Collins, London, U.K., 1967
25. Thomas, Shirley, *Men of Space* (Vol. 3), chapter "Alan B. Shepard," Chilton Company, Philadelphia and New York, 1961, pg. 206
26. Shepard, Alan B., article, "First Step to the Moon," published in *American Heritage Magazine*, July/August 1994, issue Vol. 45, No. 4
27. *Ibid*
28. *Ibid*

5

Fifteen minutes that stopped a nation

During the MR-3 countdown a number of planned communications checks had been conducted with Shepard on both UHF and HF radio. Then, two minutes prior to the planned liftoff time, the UHF radio was switched on and continuous communications were maintained between Shepard and Deke Slayton, serving as the CapCom in the Mercury Control Center. This ensured that the communications systems were fully operational at the time of launch. Shepard also received voice checks from astronauts Wally Schirra and Scott Carpenter, callsigns Chase One and Chase Two because they were circling the Cape at high level in their F-106 jets in order to follow the rapid progress of the Redstone as it ascended and headed downrange.

LIFTOFF!

Inside *Freedom 7*, Shepard noticed when the umbilical tower connection which fed power and air into the spacecraft detached. This cut direct-line connections between the blockhouse and the booster and spacecraft, which were now operating on internal power. Instead, *Freedom 7* began feeding radio telemetry information. The periscope was retracted electrically and a small door sealed the aperture. Shepard reported this, along with readings on the main bus voltage and current. "I had the feeling somehow that maybe I would've liked a little more over RF [radio frequency communications] with respect to the booster countdown steps," he later pointed out [1].

Down along the causeways and beaches, and lining the roads and highways, half a million people were present to witness history, ready to watch and wonder and shout and scream encouragement, as perfectly described in the book *Moon Shot*. "In Cocoa Beach, people left their homes to stand outside and look toward the Cape. They went to balconies and front lawns and back lawns. They stood atop cars and trucks and rooftops. They left their morning coffee and bacon and eggs in restaurants to walk outside on the street or on the sands of the beach. They left beauty parlors and barber shops with sheets around their bodies. Policemen stopped their cars and got out, the

140 **Fifteen minutes that stopped a nation**

As dawn gathered on 5 May, the media once again stood ready to record the launch of *Freedom 7*. (Photo: NASA)

In the Pad 5 blockhouse Wernher von Braun (center, wearing glasses) prepares to watch the launch. (Photo: NASA)

better to see and hear. Along the water, the surfers ceased their pursuit of the waves and stood, transfixed, swept up in the fleeting moments." [2]

During the final minute, Shepard recited to himself, "Deke and the man upstairs will watch over me. So don't screw up, Shepard. Don't screw up. Your ass is hauling what's left of your country's man-in-space program!" He was reasonably calm as the count approached zero. His left hand automatically closed over the single-twist abort handle and he kept his right hand free, ready to start the clock on the control panel.

As the countdown passed into the final ten seconds, Slayton's calm, professional countdown was accompanied by a little vibration as the Redstone's internal pumps burst into life.

"T-minus seven," Slayton intoned.

"Six, five …"

Shepard instinctively pushed his feet firmly against the capsule's interior, bracing himself for the launch. He reached up and pressed a 'ready' button that illuminated a light on Flight Director Kraft's console over in the Mercury Control Center.

"Four, three, two …"

He was conscious of his left hand gripping the abort handle. The escape tower's pyrotechnics were armed and ready in case he had to suddenly and explosively tear *Freedom 7* from the Redstone and a potential pad catastrophe.

Then, suddenly, it was T minus zero and time to go. This was the moment of truth for Alan Shepard. Two years of training had culminated in the naval aviator being strapped into a cramped capsule atop a modified missile and on the verge of making history.

At 9:34 a.m., he heard Slayton's cry of "Ignition!"

Then he was absorbed by the stupendous task at hand. "I remember hearing [the] firing command, but it may very well be that Deke was giving me other sequences over RF prior to main stage and liftoff [so] I did not hear them. I may have been just a little too excited." [3]

Within the thick walls of the Pad 5 blockhouse, former Peenemünde engineer Dr. Kurt Debus was directing the countdown along with Wernher von Braun, surrounded by members of the firing team. Although no one was directly responsible for pushing a button to launch the Redstone rocket, two members of the team had to commit to crucial roles. First was John ('Jack') Humphrey, who was responsible for pressing a launch sequencing button that issued the firing command. Then there was another of the Peenemünde engineers, Albert Zeiler. His critical, principal task was to watch the foot of the Redstone at the moment of ignition with his finger poised above an abort button. If he saw anything untoward in the color and shape of the exhaust issuing from the booster, and sensed the possibility of danger, he could press the button and instantly shut down the launch. With nothing amiss, Albert Zeiler gratefully moved his finger away from the dreaded abort button.

On board *Freedom 7*, Shepard's heart rate had temporarily shot up to 120 beats a minute. "Rumbling far below," he recalled later on. "Pumps spinning, fuel gushing through lines, joining in the combustion chamber. Before I could think about what came next, a dull roar boomed through the Redstone, rushed into the spacecraft and shook it with a surprisingly gentle touch. Thunder grew, louder and louder. 'Liftoff!' Deke called. I felt movement.

The moment of ignition, as mission MR-3 gets under way. (Photo: NASA)

"At liftoff I started a clock-timer and prepared for noise and vibration. The time-zero relays closed properly, the on board clock started properly, and I must say the liftoff was a whole lot smoother than I expected. Again I readied myself for vibration and shock. In anticipation, I'd already turned up the volume of the headphones. I didn't want to miss a word from Deke because of the still-increasing noise.

"*Freedom 7* swayed slightly. My heart pounded.

"'You're on your way, José!' Deke shouted." [4]

With the mobile launch gantry in the background, the Redstone rocket thunders into the sky. (Photo: NASA)

On the beaches and roadsides and every other possible vantage point, the throngs who were there that historic morning had moments earlier been shielding their eyes from the glare of the rising Sun. They now stood transfixed – almost stunned – as the Redstone slowly rose off the pad. There were loud cheers, shouts of encouragement and applause. When a loud crackling thunder swept across the Cape, the cheers grew ever louder as Alan Shepard was launched into the sky on board a spacecraft named *Freedom 7*.

FIRST TO FLY

"There was a lot less vibration and noise rumble than I had expected," Shepard later explained. "It was extremely smooth – a subtle, gentle, gradual rise off the ground. There was nothing rough or abrupt about it. But there was no question that I was going, either. I could see it on the instruments, hear it on the headphones, feel it all around me." [5]

Mildly surprised by the lack of vibration, Shepard was also pleased to find that he did not have to turn his radio receiver up to full volume in order to hear incoming transmissions. After communications were verified, he transmitted every 30 seconds in order to maintain voice contact and report the state of the spacecraft systems to the ground.

According to Flight Director Chris Kraft, "A communication procedure had been developed between the astronaut and the control center so that if the cabin and suit pressures were not maintained, an abort was to be initiated." This would restrict the peak altitude to 70,000 feet. "By aborting at this time (i.e., between T+1 min. 16 secs. and T+1 min. 29 secs.), the time above 50,000 feet could be limited to about 60 to 70 seconds." [6] But things proceeded smoothly.

Shepard later told *Life* magazine, "For the first minute, the ride continued smooth and my main job was to keep the people on the ground as relaxed and informed as I could. I reported that everything was functioning perfectly, that all the systems were working, that the g's were mounting slightly [just] as predicted. The long hours of rehearsal had helped. It was almost as if I had been there before. It was enormously strange and exciting, but my earlier practice gave the whole thing a comfortable air of familiarity. [And] Deke's clear transmissions in my headphones reassured me still more." [7]

The first critical moment was 1 minute 24 seconds after liftoff, when the vehicle passed through the point of maximum dynamic pressure, known in NASA parlance as Max Q, when the aerodynamic stress reached its peak. Shepard's head began to shake in an involuntary reaction to the vibration and his vision blurred a little.

"I was at two and a half times my normal weight. So far the flight was a piece of cake," Shepard later stated. "I was through the smoothest part of powered ascent, and now came the rutted road, the barrier I had to cross before leaving the atmosphere behind. [The] Redstone was hammering at shock waves gathering stubbornly before its passage, slicing from below the speed of sound through the barrier to supersonic [heading] straight up. Now I was in Max Q, the zone of maximum dynamic pressure where the forces of flight and ascent challenged the booster rocket. My helmet slammed against the contour couch. Eighteen inches before me the instrument panel

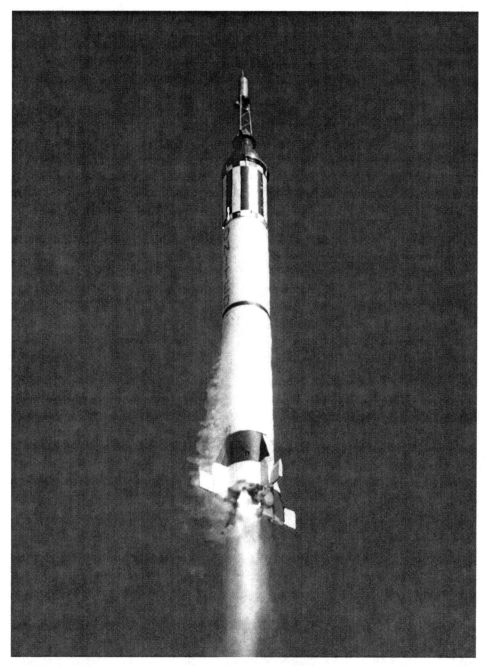

Climbing ever higher into the blue sky, Shepard prepares himself for the unsettling onset of maximum dynamic pressure, known as Max Q. (Photo: NASA)

became a blur, almost impossible to read. One thousand pounds of pressure for every square foot of *Freedom 7* was trying to crack the capsule. I started to call Deke, but changed my mind. A garbled transmission at this point could send Mercury Control into a flap. It might even trigger an abort. And then the Redstone slipped through the hammering blows into smoothness. Out of Max Q, I keyed the mike.

"'Okay, it's a lot smoother now. A lot smoother.'
"'Roger,' said Deke." [8]

The shutdown of the booster came at T+2 minutes 22 seconds at an acceleration of 6.2 g's, which meant in effect that Shepard now weighed 1,000 pounds. He was finding it difficult to talk as the g-forces constricted his throat and vocal cords. At the same time, a signal was transmitted to the spacecraft for its escape tower to separate. Above Shepard, a large solid-fuel rocket roared into life and fierce flames erupted from its three canted nozzles, ripping the tower loose from the spacecraft and pulling it away at a safe angle. "Immediately I noticed the noise in tower jettisoning. I didn't notice any smoke coming by the porthole as I'd expected I might in my peripheral vision. I think maybe I was riveted on the 'tower jettison' green light which looked so good in the capsule." [9] He promptly threw the 'retro-jettison' switch to its 'disarm' setting.

Vice President Lyndon B. Johnson watches the progress of the flight, together with President John F. Kennedy and First Lady, Jackie Kennedy. (Photo: NASA)

Ten seconds after the tower departed, the spacecraft separated from the Redstone by severing the connecting Marman clamp and firing the three posigrade rockets on the retropack for a duration of one second.

After the flight, Shepard said he was aware of the noise of the separation rockets firing. "I don't recall thinking anything in particular at separation, but there's good medical evidence that I was concerned about it at the time. My pulse rate reached its peak here [at] 132, and started down afterward." [10]

If the automatic systems had failed, the escape tower and spacecraft separation events could have been manually initiated.

"Cap sep is green," Shepard reported, as he slipped into a weightless state. As he later observed, "Moments before, I had weighed 1,000 pounds. Now a feather on the surface of the Earth weighed more than I did. Being weightless was … wonderful, marvelous, incredible. [It was a] miracle in comfort. The tiny capsule seemed to expand magically as pressure points vanished. No up, no down, no lying or sitting or standing. A missing washer and bits of dust drifted before my eyes. I laughed out loud. I'd expected silence at this point, with the atmosphere something far below me and no rush of wind despite so many thousands of miles an hour. No friction. No turbulence. But instead there was the murmur of *Freedom 7*, as though a brook were running mechanically through its structure. Inverters moaned, gyroscopes whirred, cooling fans had their own sound, cameras hummed, the radios crackled and emitted their tones before and after conversational exchanges. The sounds flowed together, some dull, others sharper. [It was a] miniature mechanical orchestra. I found those unexpected sounds most welcome; they meant things were working, doing, pushing, and repeating. They were the sounds of life." [11]

Five seconds after *Freedom 7* separated from the booster the periscope extended, and the autopilot initiated a turnaround maneuver in which the spacecraft was yawed through 180 degrees to position the heat shield forward, in the direction of reentry. In effect, Shepard was flying backwards.

One major objective of the mission – which would greatly distinguish it from the automated flight of Yuri Gagarin – was timed to start at T+3 minutes 10 seconds, when Shepard switched off the automatic control systems and took manual control of the spacecraft's attitude or angular position.

"I made this manipulation one axis at a time, switching to pitch, yaw, and roll in that order until I had full control of the craft. I used the instruments first and then the periscope as reference controls. The reaction of the spacecraft was very much like that obtained in the air-bearing [ALFA] trainer …. The spacecraft movement was smooth and could be controlled precisely." [12]

He was to maintain manual control of the spacecraft throughout the remainder of the flight by using various combinations of the attitude and rate-control systems, also known as the fly-by-wire mode.

At T+3 minutes 50 seconds, he made a number of visual observations using the periscope. These included such things as weather fronts, cloud coverage, and certain preselected reference points on the ground. As he said later, "I was zinging along high above the planet's atmosphere at better than five thousand miles per hour, but there was nothing by which to judge speed. You need relative comparison for that: a tree, a building, a passing spacecraft. My view of the outside universe was restricted to the

148 Fifteen minutes that stopped a nation

Shepard's helmeted face was filmed during the flight to record his eyes roving over the instrument panel in order to assess whether better placement of some instruments might be beneficial for future astronauts. (Photo: NASA)

capsule's two small portholes, and through those I saw that very deep blue, almost jet black, sky. There was only one available reference to tell me I was actually moving: the Earth below." [13]

He quickly realized there was a problem with the periscope. While sitting on the launch pad enduring the numerous delays he had tried to look downward through the periscope and found that he was almost blinded by sunshine filling the cabin. He had immediately inserted filters to cut down the glare, but had forgotten to remove the filters prior to launch. Now, peering through the scope, he could only see the view below in shades of gray. As he reached for the filter knob the pressure gauge on his left wrist accidentally bumped against the abort handle.

"I stopped that movement real quick," he explained later. "Sure, the escape tower was gone, and hitting the abort handle might not have caused any great bother, but this was still a test flight, and I wasn't about to play guessing games."

Gray or not, he found the view quite enthralling.

"On the periscope," he informed Mission Control. "What a beautiful view!" [14]

THE VIEW FROM SPACE

In the book, *We Seven*, Shepard related his observations of the planet passing below:

> My exclamation back to Deke about the "beautiful sight" was completely spontaneous. It was breath-taking. To the south I could see where the cloud

One of a small number of Earth observation photos taken by Shepard during his brief flight. (Photo: NASA)

cover stopped at about Fort Lauderdale, and that the weather was clear all the way down past the Florida Keys. To the north I could see up the coast of the Carolinas to where the clouds just obscured Cape Hatteras. Across Florida to the west I could spot Lake Okeechobee, Tampa Bay, and even Pensacola. Because there were some scattered clouds far beneath me I was not able to see some of the Bahama Islands that I had been briefed to look for. So I shifted to an open area and identified Andros Island and Bimini. The colors around these ocean islands were brilliantly clear, and I could see sharp variations between the blue of blue water and the light green of the shoal areas near the reefs. It was really stunning.

But I did not just admire the view. I found that I could actually use it to help keep the capsule in the proper attitude. By looking through the periscope and focusing down on Cape Canaveral as the zero reference point for the yaw control axis, I discovered that this system would provide a fine backup in case the instruments and the autopilot happened to go out together on some future flight. It was good to know that we could count on handling the capsule this extra way – provided, of course, that we had a clear view and knew exactly what we were looking at. Fortunately, I could look back and see the Cape very clearly. It was a fine reference [15].

Years later, Wally Schirra told interviewer Francis French that Shepard's remarks on his "beautiful view" were exaggerated due to his problems with the periscope. As Schirra explained, all the early astronauts felt they had some sort of obligation to say something nice about the view from space for public and press consumption. "It was just the game that people play. I'll never forget alan Shepard, on the first manned American flight, saying something to the effect of 'What a beautiful view.' I asked him later, did you see anything at all? He said 'I couldn't see a damn thing through that periscope – but I had to say something nice!'" [16]

At 5 minutes 11 seconds into the flight, *Freedom 7* reached the highest point of its ballistic arc at 115.696 miles. It now began its downward curve on a trajectory that was calculated to end with a splashdown somewhere near the naval recovery ships standing by in the waters near Grand Bahama Island, southeast of the Cape.

Deke Slayton began to recite the countdown for the retro-fire maneuver. Shepard used the manual control stick to point the spacecraft's blunt end 34 degrees below the horizon in pitch and set both the yaw and roll angles to zero.

"I worked the controls to the proper angle to test fire the three retro rockets. They weren't necessary for descent on this suborbital, up-and-down mission, but they had to be proven for orbital flights to follow, when they would be critical to decelerate Mercury spaceships from orbital speed to initiate their return to Earth.

"'Retro one.' The first rocket fired and shoved me back against my couch. 'Very smooth.'
"'Roger, roger,' from Deke.
"'Retro two.' Another blast of fire, another shove.
"'Retro three. All three retro have fired.'
"'All fired on the button,' Deke said with satisfaction." [17]

Each retrorocket was to burn for approximately 10 seconds, and they were fired in sequence at five-second intervals. "There was just a small, upsetting motion as our speed was slowed and I was pushed back into the couch a bit. But, as the rockets fired in sequence, each pushing the capsule somewhat off its proper angle, I brought it back. Perhaps the most encouraging product of the trip was the way I was able to stay on top of the flight by using manual controls." [18]

One minute after the last rocket fired, the package, its job done, blew off at T+6 minutes 14 seconds. Shepard felt the package jettison, and as he watched through the periscope he saw the straps that had held it in place begin to fall away. A green lamp was meant to illuminate on the instrument panel to indicate a successful jettison, but

As shown in this 1960 photo taken during testing at the Lewis Research Center, the spacecraft had a six-rocket retro-package affixed to the heat shield on its base. Three were posigrade rockets used to separate the capsule from the booster, and three were larger retrograde rockets to slow the capsule for reentry into the atmosphere. (Photo: NASA)

it failed to come on – the only signal failure of the entire mission. Knowing that the package had jettisoned, Shepard punched an override button and the light instantly illuminated.

Shepard verified *Freedom 7*'s HF radio, and then at T+6 minutes 20 seconds he orientated the vehicle into the reentry attitude with its blunt end 40 degrees below the local horizontal. Twenty-four seconds later the periscope retracted automatically.

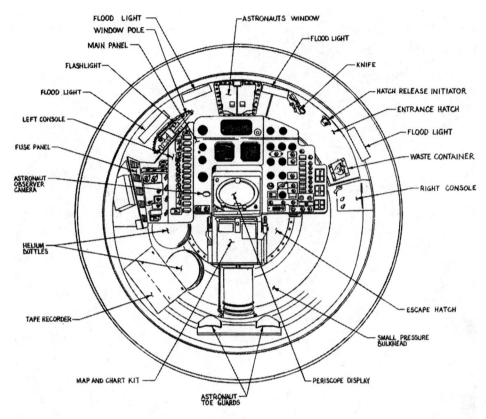

A schematic of *Freedom 7*'s instrument panel. (Photo: NASA)

The autopilot control function now allowed Shepard the freedom to conduct other flight-related functions. This included looking out through both portholes in the hope of gaining a general look at any stars or planets that might be visible, in addition to oblique views of the horizon. However, due to the Sun angle and light levels he was unable to see any celestial bodies.

END OF WEIGHTLESSNESS

At 230,000 feet, as *Freedom 7* began to penetrate the fringes of the atmosphere, a relay was actuated in response to the onset of 0.05 g. As Shepard later explained, this indicated that the reentry phase had truly begun:

> I had planned to be on manual control when this happened and run off a few more tests with my hand controls before we penetrated too deeply into the atmosphere. But the g-forces had built up before I was ready for them, and I was a

few seconds behind. I was fairly busy for a moment running around the cockpit with my hands, changing from the autopilot to manual controls, and I managed to get in only a few more corrections in attitude. Then the pressure of the air we were coming into began to overcome the force of the control jets and it was no longer possible to make the capsule respond. Fortunately, we were in good shape, and I had nothing to worry about so far as the capsule's attitude was concerned. I knew, however, that the ride down was not one most people would want to try in an amusement park [19].

It was never widely reported, but there was a little high-drama occurring at that time away from *Freedom 7*. As revealed by Shepard in an interview with *American Heritage Magazine* in 1994, it began when Slayton cautiously asked Shepard if he could see the Redstone rocket. Some engineers had expressed concern that when he fired the retrorockets and slowed the spacecraft, the tumbling booster might actually catch up. He responded in the negative, but reasoned that the booster ought, by then, to be well below his altitude. And this was indeed the case. As the booster penetrated into the atmosphere it began to disintegrate. However, as Shepard related, there was an unexpected near-miss. As the charred remains hurtled towards the ocean, sending violent shock waves through the air, this caused mounting terror for the crew of a freighter who saw a long object falling towards them. As they watched, the Redstone passed high over the ship and smashed heavily into the Atlantic just a few miles east of their position. The ship's radio operator sent out an urgent distress call, the crew suspecting they might have witnessed the death plunge of an airplane. Fortunately, a radio engineer from NBC was on Grand Bahama Island that day, heard their call, and reassured the freighter's crew that instead of a tragedy, they had witnessed the final moments of the rocket which carried America's first astronaut into space [20].

Aboard *Freedom 7*, the build-up of gravity came swiftly as the spacecraft plunged through the atmosphere. Pressed ever harder into his contour couch, Shepard noted three, then six, then nine times the force of gravity. The load peaked at 11 g's, which meant in Earth terms that he weighed close to a ton. "But I'd pulled eleven-g loads in the centrifuge, and I knew I could keep on working now." [21]

Shepard never reached the point – as he often had during grueling hours spent on the Johnsville centrifuge – of having to exert the maximum effort simply to speak or even to breathe:

All the way down, as the altimeter spun through mile after mile of descent, I kept grunting out 'O.K., O.K., O.K.,' just to show them back in the Control Center how I was doing. The periscope had come back in automatically before the reentry started. And there was nothing for me to do now but just wait for the final act to begin.

All through this period of falling, the capsule rolled around very slowly in a counterclockwise direction, spinning at a rate of about 10 degrees per second around its long axis. This was programmed to even out the heat and it did not bother me. Neither did the sudden rise in temperature as the friction of the air began to build up outside the capsule. The temperature climbed to 1,230 degrees Fahrenheit on the outer walls. But it never went above 100 degrees in the cabin or above 82 degrees in my suit [22].

Then, as the g-forces began to diminish at around 80,000 feet, Shepard switched from fly-by-wire mode back to autopilot. The altimeter was rapidly winding down, and showing 31,000 feet when Slayton's voice assured Shepard that his impact site would be right on the money.

"Great news," Shepard would later recall. "Flight computations were as close to perfect as could be, and so were the performances of the Redstone and the spacecraft …. The Cape lay 300 miles to the northwest and with the diminishing altitude would soon be out of radio contact. I signed off with Deke, telling him I was going to the new frequency.

"'Roger, Seven, read you switching to GBI [Grand Bahama Island].'

"He was eager to get the hell out of Mercury Control Center as fast as he could. I knew Gus would be right there with him, and the two of them would clamber into a NASA jet and burn sky to GBI so they could be on the ground waiting when I was delivered by helicopter from the recovery vessel." [23]

THE FINAL HURDLE

As Shepard headed to a splashdown in the Atlantic, there were still many things that had to occur, and his concentration was now on the parachute system. As he recalled in the book, *We Seven*:

The periscope jutted out again at about 21,000 feet, and the first thing I saw against the sky as I looked through it was the little drogue chute which had popped out to stabilize my fall. So far, so good. Then, at 15,000 feet, a ventilation valve opened up on schedule to let cool fresh air come into the capsule. The next thing I had to sweat out was the big 63-foot chute, which was due to break out at 10,000 feet. If it failed to show up on schedule I could switch to a reserve chute of the same size by pulling a ring near the instrument panel. I must admit that my finger was poised right on that ring as we passed through the 10,000-foot mark. But I did not have to pull it. Looking through the periscope, I could see the antenna canister blow free on top of the capsule. Then the drogue chute went floating away, pulling the canister behind it. The canister, in turn, pulled out the bag which held the main chute and pulled *it* free. And then, all of a sudden, after this beautiful sequence, there it was – the main chute stretching out long and thin – it had not opened up yet – against the sky. But four seconds later the reefing broke free and the large orange and white canopy blossomed out above me.

It looked wonderful right from the beginning. I stared at it hard through the periscope for any signs of trouble. But it was drawing perfectly, and a glance at my rate-of-descent indicator on the panel showed that I had a good chute. It was letting me down at just the right speed, and I felt very much relieved. I'd have a nice, easy landing [24].

The final hurdle 155

This photograph records the release of *Freedom 7*'s drogue chute, with the antenna canister dangling below. (Photo: NASA)

At 1,000 feet up, Shepard could see the water clearly below. The heat shield had dropped four feet as planned, to deploy the collapsible accordion-like landing bag that was stowed between it and the capsule. This perforated bag skirt of rubberized glass fiber filled with air to help to cushion the impact with the water. It provided an additional measure of shock absorption for the astronaut. Immediately after landing the parachute would be automatically disconnected, and the capsule had sufficient buoyancy to float. The landing bag and heat shield were designed to act together like a sea anchor and keep the capsule upright.

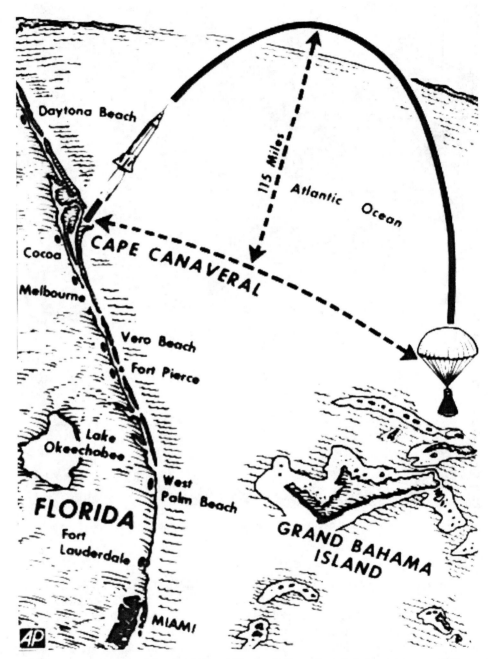

This map, prepared by Associated Press Wirephoto, shows the trajectory and ocean splashdown point for the *Freedom 7* capsule. (Drawing: AP Wirephoto)

With seconds to go before *Freedom 7* splashed in the relatively calm green water of the Atlantic, Alan Shepard braced himself for impact.

References

1. NASA Project Mercury Working Paper No. 192, *Post-launch Report for Mercury-Redstone No. 3 (MR-3)*, NASA Space Task Group, Langley Field, VA, 16 June 1961
2. Alan Shepard, Deke Slayton, Jay Barbree, Howard Benedict 1994 *Moon Shot: The Inside Story of America's Race to the Moon.* Virgin Books, London, p. 113
3. NASA Project Mercury Working Paper No. 192, *Post-launch Report for Mercury-Redstone No. 3 (MR-3)*, NASA Space Task Group, Langley Field, VA, 16 June 1961
4. Shepard, Alan B., article, "First Step to the Moon," published in *American Heritage Magazine*, July/August 1994, issue Vol. 45, No. 4
5. Carpenter, S., Cooper, Jr. L, Glenn, Jr., J., Grissom, V., Schirra, Jr., W., Shepard, Jr., A., and Slayton, D., *We Seven*, Simon and Schuster Inc., New York, NY, 1962, pg. 250
6. Kraft, Christopher C., Jr., Extract, *Flight Plan for the MR-3 Manned Flight*, from National Aeronautics and Space Administration, National Institutes of Health and National Academy of Sciences, *Results of the First U.S. Manned Suborbital Space Flight*, NASA HQ, Washington, D.C., 6 June 1961
7. Shepard, Alan, oral article, "Light This Candle," for *Life* magazine, Vol. 50, No. 20, May 19, 1961
8. Shepard, Alan B., article, "First Step to the Moon," published in *American Heritage Magazine*, July/August 1994, issue Vol. 45, No. 4
9. NASA Project Mercury Working Paper No. 192, *Post-launch Report for Mercury-Redstone No. 3 (MR-3)*, NASA Space Task Group, Langley Field, VA, 16 June 1961
10. Shepard, Alan, oral article, "Light This Candle," for *Life* magazine, Vol. 50, No. 20, May 19, 1961
11. Shepard, Alan B., article, "First Step to the Moon," published in *American Heritage Magazine*, July/August 1994, issue Vol. 45, No. 4
12. Shepard, Alan B., Jr., Extract, *Pilot's Flight Report*, from National Aeronautics and Space Administration, National Institutes of Health and National Academy of Sciences, *Results of the First U.S. Manned Suborbital Space Flight*, NASA HQ, Washington, D.C., 6 June 1961
13. Shepard, Alan B., article, "First Step to the Moon," published in *American Heritage Magazine*, July/August 1994, issue Vol. 45, No. 4
14. *Ibid*
15. Carpenter, S., Cooper, Jr. L, Glenn, Jr., J., Grissom, V., Schirra, Jr., W., Shepard, Jr., A., and Slayton, D., *We Seven*, Simon and Schuster Inc., New York, NY, 1962, pg. 254
16. Schirra, Wally, interview with Francis French, San Diego, California, 22 February 2002
17. Shepard, Alan B., article, "First Step to the Moon," published in *American Heritage Magazine*, July/August 1994, issue Vol. 45, No. 4
18. Shepard, Alan, oral article, "Light This Candle," for *Life* magazine, Vol. 50, No. 20, May 19, 1961

19. Carpenter, S., Cooper, Jr. L, Glenn, Jr., J., Grissom, V., Schirra, Jr., W., Shepard, Jr., A., and Slayton, D., *We Seven*, Simon and Schuster Inc., New York, NY, 1962, pg. 258
20. Shepard, Alan and Deke Slayton, with Jay Barbree and Howard Benedict, *Moon Shot: The Inside Story of America's Race to the Moon*, Virgin Books, London, U.K., 1994, pp.121,122
21. Shepard, Alan B., article, "First Step to the Moon," published in *American Heritage Magazine*, July/August 1994, issue Vol. 45, No. 4
22. Carpenter, S., Cooper, Jr. L, Glenn, Jr., J., Grissom, V., Schirra, Jr., W., Shepard, Jr., A., and Slayton, D., *We Seven*, Simon and Schuster Inc., New York, NY, 1962, pg. 259
23. Shepard, Alan B., article, "First Step to the Moon," published in *American Heritage Magazine*, July/August 1994, issue Vol. 45, No. 4
24. Carpenter, S., Cooper, Jr. L, Glenn, Jr., J., Grissom, V., Schirra, Jr., W., Shepard, Jr., A., and Slayton, D., *We Seven*, Simon and Schuster Inc., New York, NY, 1962, pg. 260

6

Splashdown!

It was April 1961 and 20-year-old Air Controlman 3/c (Third Class) Ed Killian from Texas had been serving aboard the *Essex*-class aircraft carrier USS *Lake Champlain* (CVS-39) for eighteen months. During that time "The Champ" – as she was fondly known to her crewmembers – had mostly been engaged in an anti-submarine patrol rotation out of NAS Quonset Point, Rhode Island. Towards the end of the month the ship finished a three-week sweep northeast of Norfolk, Virginia, and her exhausted crew were eagerly anticipating some shore leave.

Killian, days away from his 21st birthday, was looking forward to celebrating this milestone in his life in New York City. But then Capt. Ralph Weymouth, the ship's commanding officer, made an announcement over the ship's loudspeakers that left Killian's liberty plans in tatters. Unbeknownst to the crew at the time, however, this change in their schedule would give every man on board the chance to participate in, and be an eyewitness to, an historic event. The ship was to set a new course for the Mayport Naval Station near Jacksonville, Florida. Once there, and after a couple of days, they were to join other units in an area to the east of nearby Cape Canaveral for what was vaguely described as some "special ops."

"This news of extending the cruise didn't sit well with the crew in general, and there was a lot of grousing about it," Killian reflected. "The scuttlebutt was that we were going to do another anti-submarine demonstration for some bigwigs, like the one we had done the previous summer for the Latin American generalissimos and admirals in the Caribbean. Later, while we finished polishing the brass fittings on Pri-Fly's windows [the control tower for flight operations, known as Primary Flight Control], Cdr. Howard Skidmore, the ship's Air Officer, known aboard as the 'Air Boss,' came in with a big excited grin on his face and told us that the ship was to be the recovery vessel for a space shot from Cape Canaveral. We had heard about the chimpanzee Ham's flight and recovery in January, and being a recovery vessel for another 'monkey flight' held no excitement for us. Such an event was viewed as a poor trade-off for missing liberty in Quonset Point." [1]

The USS *Lake Champlain* in 1960. (Photo: U.S. Navy Naval Historical Center)

CIVILIANS ON BOARD

The crew's puzzlement grew in Mayport the day after their arrival, with dozens of civilians boarding the carrier. Killian recalls "about fifty NASA and government officials, photographers by the dozen, and boxes of equipment by the hundreds." Frustration grew amongst the crewmembers, already annoyed at not returning to Quonset Point, who now found their normal routines delayed by these civilians, while the mess hall, already small, was becoming jammed with extra bodies at meal times. Often the line for food wrapped around several frames of the third deck space, creating much grumbling and dissention.

Eventually, Cdr. Skidmore gave his small group of six air controllers in Pri-Fly additional details about the special operation for which the USS *Lake Champlain* had been selected. As the ship and her crew had performed well in fleet-wide operational competitions, she had been selected as the prime spacecraft recovery vessel for the United States' first manned space shot, then scheduled for 2 May.

The recovery task force was actually comprised of several task groups, each under an individual commander, dispersed along the projected track of the spacecraft. The task

Ed Killian on board the USS *Lake Champlain* before liberty in Charlotte Amalie, U.S. Virgin Islands, in February 1961. (Photo courtesy of Ed Killian)

group in the predicted landing area was commanded by R/Adm George P. Koch, Commander, Carrier Division 18, flying his flag aboard the USS *Lake Champlain*. A crucial element of the recovery task force was a flotilla of destroyers commanded by R/Adm Frederick V.H. Hilles. In cooperation with fellow flag officer Koch aboard the USS *Lake Champlain*, Hilles would exercise his command of the destroyers from the

Recovery Control Room located at NASA's Mercury Control Center at Cape Canaveral. The units of this group were:

Name	Type	Commander
Aircraft carrier		
USS *Lake Champlain*	CVS-39	Capt. R. Weymouth
Destroyers		
USS *Decatur*	DD-936	Cdr. A.W. McLane
USS *Wadleigh*	DD-689	LCdr. D.W. Kiley
USS *Rooks*	DD-804	Cdr. W.H. Patillo
USS *The Sullivans*	DD-537	Cdr. F.H.S. Hall
USS *Abbot*	DD-629	Cdr. R.J. Norman
USS *Newman K. Perry*	DD-883	Cdr. O.A. Roberts
Minesweepers		
USS *Ability*	MSO-519	LCdr. Larry LaRue Hawkins
USS *Notable*	MSO-460	Lt. Freeland
Salvage and recovery		
USS *Recovery*	ARS-43	LCdr. Robert Henry Taylor
Tracking ship		
USAF *Coastal Sentry*	T-AGM-50	-

Two of the six destroyers were positioned 100 miles or more from the USS *Lake Champlain*, between Cape Canaveral and the projected recovery area, but the others remained in close contact with "The Champ."

As planned, the aircraft carrier departed Mayport and sailed into position for the recovery operation about 300 miles east-southeast of the Cape. However, inclement weather forced the launch to be delayed, and there was a further delay two days later before the countdown finally picked up again on the morning of Friday, 5 May.

Because of their advantageous view from Pri-Fly, Cdr. Skidmore had arranged for those who worked there to pool their film with NASA photographer Dean Conger. A well-respected photographer for the *National Geographic* magazine, Conger was "on loan" to NASA as one of the official photographers to record the recovery operation. Appreciating the historic nature of Alan Shepard's flight, Skidmore set up extensive photographic coverage by positioning volunteer officers and enlisted men at different vantage points on the ship so that the recovery could be recorded on film. When this innovation was combined with the work of the NASA photographers, it resulted in a magnificent, sweeping coverage of the occasion.

After discussing the expected capsule retrieval with senior crewmembers of ships in the recovery Task Force, NASA Space Task Group representatives Martin Byrnes, Robert Thompson, and Charles Tynan determined that the ship's crewmen assigned to handle the spacecraft were not fully trained in the specifics of what was expected of them, so they initiated a brief education program for the crew. This included the provision of printed information sheets and screening a film on the recovery of the capsule containing chimpanzee Ham earlier that year. According to *This New Ocean: A History of Project Mercury*, published by NASA, "Tynan also carefully briefed each man charged with capsule-handling duties on his particular role." [2] The carrier's

NASA Recovery Team Leader Charles I. Tynan, Jr. (seated), briefs the USS *Lake Champlain*'s recovery team officers on spacecraft retrieval. Standing with his hands on his hips is Richard Mittauer, a NASA Headquarters Public Affairs Officer. The Executive Officer of the ship, Cdr. Landis Doner, is at center rear. (Photo courtesy of Ed Killian)

recovery team consisted primarily of enlisted crewmembers led by the Flight Deck Officer and a number of enlisted Chief Petty Officers of the Air Department; none of whom were at the Tynan briefing. Those in attendance were senior officers who then gave orders and instructions to the carrier's recovery team. Strangely, not even the Air Officer who led the Air Department was invited to the Tynan briefing.

EYEWITNESSES TO HISTORY

The day of the MR-3 launch left Ed Killian with many memories. "Air Boss Howard Skidmore was in Pri-Fly with his several cameras. I'd never seen so many observers on the island's superstructure.[1] Every exterior catwalk – every vantage point – was crammed with photographers, reporters, scientists, NASA technicians, and military representatives of all the branches. The crew appeared to be going about its regular routine. There were no more personnel visible on deck than would have been on duty for regular flight operations. Then came an announcement from the ship's 1MC [the

---

[1] The "island" of a carrier includes the command center for flight deck operations, captain's bridge, admiral's bridge, and the navigation, meteorology and signal bridges.

main circuit] loudspeaker that the astronaut was in the capsule and the launch was counting down. The announcement then said that the crew was permitted topside on the flight deck to view the [splashdown] event." They were to observe the recovery standing aft of a raised arresting gear cable, and aft of the island. Marine guards had been posted opposite the crew to keep them from moving forward prior to the arrival of the Marine helicopter bearing the spacecraft – more as a matter of safety than one of security.

"Following this progress report on the 1MC, the ship's crew suddenly started to appear on the flight deck. They literally poured out of every hatch, filled every deck catwalk and spilled onto the flight deck. Hundreds of them, in all manner of jersey colors; red for gas handlers and ordnance men; yellow for plane directors; green for electricians and aviation technicians; blue for plane spotters; brown for ship's hangar deck crews. Sailors in blue dungarees, cooks in cook-caps, the dirtiest First Division bosuns, and sailors of every discipline on the ship poured out onto the flight deck aft of the island, their eyes agog at all the activity on the forward flight deck area. Some crew members whose duty stations were on the island were able to get on the island catwalks to observe the recovery."

With the launch imminent, Marine Corps Sikorsky HUS-1 Seahorse helicopters on the deck forward of the island were prepped for liftoff, ready to proceed to the recovery site which was expected to be several miles off the carrier's port bow.

Marine Corps lieutenants Wayne Koons from Lyons, Kansas, and George Cox from Eustis, Florida, had been attached to squadron HMR(L)-262. They were aboard

The island structure of the USS *Lake Champlain* on 5 May 1961. (Photo courtesy of Ed Killian)

"The Champ" on temporary assigned duty, having been selected to fly the primary recovery helicopter. Two other Marine Corps choppers were also in the same flight, and ready to act in a backup capacity if necessary. Koons had participated in three previous at-sea retrievals; two as copilot and one as pilot. Cox had been involved in the recovery of the MR-2 capsule that carried chimpanzee Ham on 31 January, just three months earlier. About two months prior to the flight of MR-3, Koons and Cox had participated in live training with Shepard, Gus Grissom and John Glenn. In this training the egress trainer was placed in the "back river" at Langley AFB, Virginia, with an astronaut inside. Three live retrievals were made, one with each astronaut, in this calm water environment.

FROM THE LAND TO THE SEA

Wayne Koons was a farm boy from Rice County, Kansas. After leaving from high school he attended Ottawa University in Ottawa, Kansas, where he received degrees in physics and mathematics before electing to join the U.S. Marine Corps and then opting to become a helicopter pilot. He recalls one day in 1959 while stationed in MCAS New River, North Carolina, when a squadron clerk ran up to him and said the commanding officer wanted to see him urgently.

"I was literally apprehensive as I went over to the hangar. The skipper told me they had had an enquiry about using helicopters to retrieve astronauts and spacecraft from the ocean. And I kept thinking, 'What do you mean – astronauts?'" The term was new to him [3]. He later learned that he had been chosen for this task because he was the only pilot of 250 possible candidates to have a technical degree. Helicopters had been selected for the sea recovery operation because NASA's engineers were not overly confident in the seaworthiness of the Mercury capsules, while flight surgeons were not confident about the physical shape that a person might be in after making a flight in space. "They wanted to get the astronaut and the spacecraft out of the water quickly," Koons explained [4].

Lt. Koons was assigned as Project Officer to the retrieval squadron of the Space Task Group, and used his experience and technical abilities to assist in developing recovery techniques and procedures for the yet-to-be-built Mercury spacecraft. His duties included training the squadron pilots. Another part of his assignment involved the design and testing of a special reinforced loop on top of the capsule to enable the helicopter's copilot to snag it using a long pole with a curved attachment on the end known as the "shepherd's hook."

As Wayne Koons described the procedure to the author, "The shepherd's hook was attached to the lower end of a steel cable, which was engaged in the spacecraft lifting loop by the copilot using a long pole. The upper end of the cable was locked into the helicopter's cargo hook before the helicopter lifted off from the ship. Lifting the spacecraft was accomplished by raising the helicopter. Once the spacecraft was on the ship's padded skid, the helicopter cargo hook was opened, thus releasing the cable which stayed with the spacecraft." [5]

In order to get the procedure right, countless training test flights were conducted using full-size representations of the Mercury spacecraft. In one test, a "boilerplate" capsule was dropped from 1,000 feet over Chesapeake Bay in Virginia. Once it had parachuted into the water, the capsule was retrieved by the crew of a twin-engine helicopter using a "shepherd's hook" in what NASA later described as a "successful" drop.

The squadron's commanding officer informed Koons of the decision to have him serve as the lead pilot for this mission. "He asked me if I had a particular choice in copilot, and I said, 'I sure do,' because at that point I had been working regularly for several months with a copilot named George Cox. George and I just really got along well. We were kind of like twins. You know, we didn't have to say everything that we communicated. We thought alike, worked well together, and were comfortable with each other. And George was really eager to do it, enjoyed working the mission. So that was the basic set up."

Eventually Koons found out that the retrieval would take place using an aircraft carrier as the destination. "We did some flight training right after we got aboard, and the Air Boss had set the flight deck up and put the skid with the mattresses on it up close to the bow. So we went off, and one of our helicopters dropped the boilerplate into the water, and then we went and picked it up and delivered it back to the ship." When this proved a difficult operation visually for the pilot, the Air Boss rearranged his flight deck, shifting airplanes around and placing the skid on the rear of the flight deck. "So then when we tried it that way, it was much better, because I had the island in my field of view, and out the front windshield I could see the front part of the flight deck. So it made it much easier, a lot easier to maintain a good visual reference while I was setting the thing down." [6]

Along with Capt. Allen K. Daniel, Jr., Koons recovered the MR-1A capsule from the Atlantic on 19 December 1960 and safely deposited it aboard the aircraft carrier USS *Valley Forge* (CV-45).

Despite their intense training, the helicopter team often struggled with the payload capability of the aircraft balanced against the weight of the spacecraft. As Koons put it, "The dry weight of the spacecraft with astronaut was well over one ton. When the project started, the spacecraft weight was estimated to be about 1,800 pounds. As the design matured, the weight increased to the point that the complete retrieval weight for MR-3 approached 2,900 pounds. The helicopters were stripped of all unnecessary weight. The extra seats, the APU [auxiliary power unit], heater, some avionics, and the (only) life raft were removed for Mercury retrieval flights. Also, the fuel load was tailored. To explain: The downrange helicopters were tasked to go as far as 115 nautical miles from the ship to retrieve a spacecraft. To accomplish this, the fuel load was normally set for a retrieval close aboard the ship. This reduced load gave the helicopter the lift margin needed to accomplish the capsule retrieval. If the spacecraft landed some distance from the ship, once the distance was known, the fuel load was adjusted to optimize the lift capability at the spacecraft, with adequate reserves for the return. The fuel calculations assumed the ship proceeded toward the spacecraft at its best speed after the launch of the helicopters.

"This scenario happened on the MR-2 flight. Range to the spacecraft was about 100 nautical miles. A big problem arose when the ship lost boiler power and went dead in the water after the launch of the helicopters, and we were 'running on fumes' by the time we actually got back to the ship with the spacecraft." [7]

Alan Shepard discusses recovery procedures with Wayne Koons and George Cox. (Photo courtesy of Wayne Koons)

THE MAN WITH THE CAMERA

As the day of the space shot drew nearer, Koons and Cox had to confer with media people on the USS *Lake Champlain* in order to coordinate the best possible coverage of the retrieval of Alan Shepard and his spacecraft.

One person who impressed Koons was *National Geographic* photographer Dean Conger. "He was part of the pool. He had been out with us on one prior mission, I think. Dean showed up with a camera that he asked to clamp onto the side of the helicopter, where it would be looking down as we did the retrieval. It had a wide-angle lens. I can't remember how many exposures he said it had. It must have been just a standard thirty-six-exposure roll. But he said it was automatic, and he could set it up to take just one shot per second ... it was actually on one of the little struts that held the [personnel] hoist. And the other thing, if we could just remember to turn it on when we started doing the pickup." [8]

On loan to NASA, *National Geographic* photographer Dean Conger attaches an automatic camera to Marine helicopter #44's winch-hoist frame assisted by a Marine corporal from HMR(L)-262. (Photo courtesy of Ed Killian)

As Dean Conger recalled for the author, "Logically I wanted to be on the prime pickup chopper, but that was ruled out for weight reasons. They flew with only the pilot and copilot. When it came time to pick up the astronaut the copilot would leave the cockpit and go to the doorway to operate the winch.

"On another ship for an earlier flight [MR-2 with Ham] I talked with the Marine crew chief – unfortunately, I don't remember his name – who was extremely helpful with my idea of placing a remote camera somewhere. He said that he could make a bracket. He welded together a bracket out of 2-inch strap iron which had a pocket to accept a Nikon fitted with a 250-exposure back, and we would run a cord down to a battery pack which was fastened to the side of the cabin door. I believe it was taped there. The Nikon battery pack only had a push button. The chief made a slip-on clip that the copilot would push in order to hold the button down for continuous shooting. It was all very crude, but it worked.

"The flight plan was as follows: When the capsule landed, a long antenna would deploy straight up. So the first maneuver was to fly in and the copilot, having left the cockpit, would snip off the antenna using an explosive bolt cutter on a long pole so that it wouldn't interfere with the rotor blades. Then the chopper would circle back, stabilize the capsule, and lift the astronaut up. In testing, this had checked out to take about 10 minutes. In their flight plan, the copilot would push the camera switch as they approached to cut the antenna. The problem for me was that the film would run out in about 2.5 minutes. A technician on the ship said he would wire a resistor into the cord to slow down the camera. After the fact, Nikon said it ought not to have worked at all!" [9]

It was something of a gamble, but Dean Conger was an experienced, professional photographer determined that this particular day in history should be recorded for posterity with only the very best images.

CALLING *FREEDOM 7*

At long last the 1MC on the USS *Lake Champlain* announced that *Freedom 7* had been launched successfully, and the sailors were told that if they looked to the west in several minutes they should be able to witness the spacecraft's return.

At about the nine-minute point in the flight, Capt. Weymouth called Pri-Fly, and Ed Killian's personal involvement in the MR-3 mission started. Weymouth said the NASA people on board the ship had not been able to reach the astronaut by radio. As Killian recalled, "The capsule carried both UHF and VHF radio units, but the VHF did not seem to be working. It was suspected that reentry ionization had screwed up radio communications such that NASA was unable to contact Shepard, but there was a scary possibility that something had gone seriously wrong. The Captain asked the Air Boss to see if Pri-Fly could raise the capsule on their UHF radio. We hadn't been informed as to the capsule's call sign, since we weren't expected to be involved in communicating with Shepard, so we didn't know that his call sign was '*Freedom 7*.' The Air Boss called 'Mercury' several times on the VHF radio without success, and then asked me to try the AR-15. This UHF aircraft radio was not standard equipment for Pri-Fly; it had been added several months earlier as a backup to the VHS system routinely used to contact the aircraft under control of Pri-Fly. As senior controller on my shift, I generally wore the mike and headset, and operated the AR-15. It was not connected to any loudspeaker in Pri-Fly, so those communications were more or less private. Having failed to reach Shepard on VHF, the Air Boss turned and told me to call Shepard."

Killian then began his attempt to reach the spacecraft. "*Mercury, Mercury*, this is *Nighthawk*. Do you read? Over." There was no response. He repeated the entire call several times without success, and then told Air Boss Howard Skidmore he'd had no luck. Told to keep trying, he repeated the call several times more. He could see the concern on Skidmore's face, and when he glanced up at the bridge he could also see Capt. Weymouth looking down at him with an equally worried expression.

Seconds later, an urgent call came over the intercom. "Pri-Fly, this is the Captain. Have you been able to raise Mercury?" Replying in the negative, Skidmore motioned for Killian to keep trying. As he repeated the call, the AR-15 sputtered and crackled, then cleared up and an exuberant voice responded with, "Roger, *Nighthawk*, this is *Mercury*. Boy, what a f...ing ride! Ho-lee s...! Goddam, that was something!" It was Alan Shepard himself.

"His excited reply startled me," Killian says, "and such language wasn't common in the radio traffic monitored in Pri-Fly. I was a little embarrassed, but pretty sure no one else had heard his outburst." Happily for newspaper editors the following day, Shepard repeated his excited catch-cry of "Boy, what a ride!" to Capt. Weymouth on arriving on the ship – minus the profanity – which became the headline tag on many a global newspaper that day.

As Killian continued, "Shepard's voice was at a high pitch in his excitement and it was obvious that he was glad to have made contact with someone on Earth after his short, but explosive, sojourn into space. We all knew that several previous launch attempts had exploded on the pad or shortly thereafter; he must have felt exhilarated just to be alive. His excited and happy response told me all we had wanted to know. I turned to my right and looked up to Capt. Weymouth on the bridge, and with a big smile on my face I gave him the thumb-and-forefinger 'OK' sign. Alan Shepard, America's first astronaut, was A-OK. I made contact with him again, confirming that we'd received his transmission. 'Roger, ride, *Mercury*,' I said, '*Nighthawk* has a visual on you now, four miles off our port bow. We are making for your location and the choppers are airborne for your recovery. Do you read? Over.'

"I could tell from the tone of his voice when he responded that he was relieved to know that we were so close by. He regained some composure and acknowledged my transmission, thanking me for the information. I was watching the capsule swinging gently below the parachute just a few points off our port bow as I spoke to Shepard on the AR-15. Finally, NASA's VHF radio finally broke in and I dropped out of the communications loop."

These days Ed Killian is still amused at the blue language used by Shepard after his fiery reentry, and coyly wonders whether it ought perhaps to have been reported correctly. "It was truly memorable," he notes, "but the language was not scripted and it was just not acceptable for public audiences. Later, the astronauts would become more adept with 'politically correct' language. For now, Shepard had been honest in his reaction to the historic, and patently dangerous, personal experience."

Whereas nowadays nearly every word of an American space mission is generally broadcast live, back then NASA was far less inclined to allow members of the public to listen to uncensored recordings of the Mercury astronauts under stress or even in danger. Only those who were fully involved in the missions had a "need to know" insight into the actual spacecraft-to-ground transmissions. Every word between the astronauts and the ground was filtered through Lt. Col. John ('Shorty') Powers, the space agency's Public Affairs Officer, who then released a sanitized version of what had been said to the public. Even post-flight films had an edited voice track of the astronaut involved in that mission, which became the "official" NASA version of the flight. It is therefore hardly surprising that Shepard's excited expletives, normally quite unremarkable for test pilots, were censored and never made public.

A HERO RETURNS

At last, and for the first time in history, a spacecraft had completely returned to Earth from space with a human being on board. Even Yuri Gagarin and his Soviet space masters could not match this technological feat – although several years would pass before the truth was finally revealed – as the first cosmonaut had been ejected from his descending Vostok capsule to parachute to the ground according to highly secret plans. A workable and reliable retrorocket system had yet to be developed by Soviet engineers that would allow a controlled, soft landing for returning cosmonauts. As a result, all six Vostok pilots were required to eject from their vehicles. One reason for keeping this secret was that the Fédération Aéronautique Internationale, which rules on aviation records, requires that a pilot be in his aircraft at the time of landing.

Arthur Cohen, who headed the IBM team that created and ran the computers that modeled and then tracked the flight, would report, "Everything went smoothly. All the plotting was perfect. Everything on the flight was nominal. There was really no problem whatsoever. It landed at exactly the right place." [10]

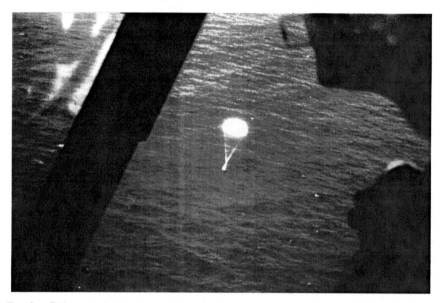

Freedom 7 photographed moments before it splashed down in the Atlantic. (Photo courtesy of Ed Killian)

The capsule hits the water, concluding America's first manned space mission. (Photo courtesy of Ed Killian)

For his part, Shepard was exuberant over his safe return. "Splashdown!" he later wrote of that moment. "Into the water we went with a good pop! Abrupt, but not bad. No worse than the kick in the butt when I was catapulted off a carrier deck. This was home plate!" [11]

Prior to impacting the water, he had removed his knee straps. He now began his post-splashdown procedures. The first thing was to release his face plate seal bottle. This was a small pressure bottle joined by a thin hose to a connector next to his left jaw. It was used to pressurize a pneumatic seal when the face plate was closed. He then removed the exhaust hose from the helmet.

Once *Freedom 7* had splashed down, it quickly swayed over on its side, about 60 degrees from vertical, covering the right porthole with seawater and causing Shepard to lean over onto his right side in the couch. Seeing that the porthole's exterior was completely under water, he deduced that everything was going to plan. Through the other porthole he could also make out the bright green fluorescing dye automatically spreading out through the water to mark his position for recovery aircraft.

"As I waited for the shifted balance to right my great spacecraft – but lousy boat – I kept thinking about the chimp's near-disappearance beneath the ocean. I checked and checked the cabin for leaks, ready to punch out. But I stayed dry."

Shepard then activated the 'rescue aids' switch to jettison the reserve parachute, thereby reducing some of the top-heavy weight of the spacecraft and allowing it to stabilize itself upright in the water. "Shifting the center of gravity had worked, and the capsule came back upright." [12] Flipping the switch also released the HF antenna, although he did leave his transmit switch in the UHF position. To his relief, all of the recovery aids seemed to be working well, although he was not to know that the HF antenna had failed to extend skywards. However, with the recovery ships and aircraft in the immediate vicinity its function as a location device was not needed.

"I'd broken my helmet at the neck ring seal at this point, and I did no transmitting here," he later observed in his initial flight report aboard the recovery carrier. "I left the switch on R/T [receive/transmit] because I didn't want any discharge from the UHF antenna [13].

"I could not see any water seeping into the capsule, but I could hear all kinds of gurgling sounds around me, so I wasn't sure whether we were leaking. I remember reassuring myself that I had practiced getting out of the capsule under water and that I could do it now if I had to. But I didn't have to try. Slowly, but steadily, the capsule began to right itself." [14]

It was time to assure everyone he had survived the splashdown by making contact with the communications relay airplane (codename *Cardfile Two Three*) which was circling overhead.

> "*Cardfile Two Three*," he called. "This is *Freedom 7*. Would you please relay: All is okay."
> "This is *Two Three*," came the reply. "Roger that."
> "This is *Seven*. Dye marker is out. Everything is okay. Ready for recovery."
> "*Seven*, this is *Two Three*. *Rescue One* will be at your location momentarily."

Shepard continued his preparations to leave *Freedom 7* as the spacecraft became almost vertical in the water. He began to document the instruments prior to shutting down the power. "I had just started to make a final reading on all of the instruments when the helicopter pilot called me. I had already told him that I was in good shape, but he seemed in a hurry to get me out. I heard the shepherd's hook catch hold of the top of the capsule, and then the pilot called again." [15]

RIGHT ON THE SPOT

Normally, the helicopter carried three crewmembers including a crew chief, but due to the expectedly high weight of the spacecraft the crew had been reduced to just the pilot and copilot. Wayne Koons revealed he could hear Shepard's transmissions from about the time *Freedom 7* reached 85,000 feet, when the astronaut came within range of their receiver. "And then we were actually talking with him after he had struck the water and was waiting to be picked up. We were right on the spot. We were waiting for [the capsule] to hit. We were circling the parachute as it came down." [16]

At the time of splashdown, George Cox was at the flight controls with Koons. All seemed to go as expected once the capsule was in the water, so Cox left his cockpit seat and shinnied down below to make ready for the task of retrieving the astronaut and his vessel. First they had to verify that the parachute had been released from the bobbing craft by Shepard and had sunk beneath the surface, which, as Koons stated, was "something we always had to watch out for, because if there was any part of that chute above water, you ran the chance of the rotor wash picking it up and inflating it again. So we had to be sure it was off and sunk in the water so that it wasn't going to come up." [17]

Koons and Cox kept waiting for the spacecraft's long HF antenna to pop up, but when they couldn't see it Koons moved into position above *Freedom 7*. Ordinarily, one of Cox's immediate duties would have been to lower a tool that had a bolt-cutter with explosive squibs at its end to sever the antenna to prevent it interfering with the raising of the astronaut into the helicopter. To his surprise, he found that the squibs were absent, but since the antenna had not deployed he re-stowed the antenna cutter. While Koons skillfully hovered above *Freedom 7*, Cox used the "shepherd's hook" to snag the recovery loop on top of the spacecraft. At that point, without warning, the HF antenna telescoped upwards and its tip struck the helicopter's fuselage.

According to the post-flight report which NASA declassified in June 1973, "The explosively actuated telescoping HF recovery antenna [was] erected after helicopter hook-onto the capsule but prior to pilot egress. The activation time was normal; the helicopter moved into recovery position earlier than planned. The helicopter pilot observed only a 10-foot length of the antenna rather than the normal 16 feet. Later inspection showed the last 6 feet had probably been blown off at erection. There was no evidence of this section striking the helicopter. The remaining 10-foot section was not cut off by the helicopter crew, and caused no difficulty in recovery." [18]

Koons would later say that the recovery was otherwise a fairly routine operation. "The only anomaly we had, was that that antenna did pop up sometime. I'm not sure when it did, but we found a dent in the bottom of our helicopter …. But I never knew when that happened, when it finally decided to go." [19]

Marine helicopter #44 moves into position low above *Freedom 7*. Seconds later the HF antenna deployed, striking the hovering craft's fuselage. (Photo: U.S. Navy)

Charles Tynan, the senior NASA representative present, told the author about the missing explosive squibs. "The squibs were for the tool the helicopter crewman was to use to cut off the HF antenna, because it was long enough to contact the helicopter rotors. The Marine helicopter mechanic's tool box was broken into the night before the recovery and all the pyrotechnic squibs were stolen. There was plenty of time for more squibs to be flown out to the carrier from the Cape, but the Captain would not let this happen because he didn't want the bad publicity for his command." [20]

When everything was ready, Cox prepared to hoist Shepard up into the helicopter. "I was in the belly of the aircraft and operated the hoist [which] took him from the capsule up to the cabin of the helicopter. [We] hooked onto the capsule and started pulling it up to steady it upright in the water. We told Commander Shepard we were ready for him to come out and recover him, and he asked us to raise the capsule a little bit higher." [21]

Shepard later said this was because he could still see water out of a porthole and wanted to avoid getting any of it inside the spacecraft. "I'm not sure he heard me at first, but I was able to get through to him that I'd be coming out as soon as he lifted the door [hatch] clear of the water." For this first flight of the Mercury spacecraft, *Freedom 7* possessed a mechanical hatch which was fitted with latches that were to be actuated by a handle that Shepard would crank. But first Shepard had to attach a metal cable to the hatch in order to prevent it from being lost once free. "I called the helo and told him I was ready to come out, and he verified that he was pulling me up. I told him I was powering down and disconnecting communications." [22]

Koons obliged the request by raising the capsule a foot or two higher. "We were all very aware that the spacecraft hatch was normally partially below the waterline," he says. "We knew for sure that opening the hatch too soon would result in flooding the spacecraft, so Shepard's request to raise the spacecraft higher was redundant in that we were in the process of doing just that." [23]

Shepard now said he would be out in about 30 seconds. By pre-arrangement, if he had decided at this point not to exit the capsule then Koons would have hoisted it out of the water and transported him to the carrier inside the spacecraft. Having opted to egress, all Shepard had to do was to rotate the locking handle so that the hatch would detach and then scramble out.

"The door was ready to go off. I disconnected the biomedical packs. I undid my lap belt, disconnected the communications lead, and opened the door." As the hatch opened Shepard allowed it to fall away. Unfortunately, even though he had properly affixed the hatch to the capsule by the cable, the crimped metal clip on the lower end of the cable had been closed over the plastic sheath instead of over the cable, and this allowed the cable to pull out. The unrestrained hatch plunged into the water and sank to the ocean floor.

As Shepard later said, he climbed out "and very easily worked my way up into a sitting position on the door sill. Just prior to doing this, I took my helmet off and laid it over in the position in the ... as a matter of fact, I put it over the hand controller." [24] He began looking upwards for the "horse collar" recovery harness, which Cox was in the process of lowering to him.

"It went like another practice run," Cox pointed out later. "In just a moment we began the hoist. He was giving me a big grin all the way up, and a big thumbs-up. He looked like the same Commander Shepard that I'd known before and worked with,

Lt. George Cox winches Alan Shepard into the helicopter. (Photos courtesy of Dean Conger/NASA)

except a little happier than before." [25] As he ascended, Shepard brushed up against the truncated antenna, but it was flexible and no harm was done.

As requested, Koons had reminded Cox to turn on photographer Dean Conger's camera before hoisting the astronaut, and received confirmation this had been done. "Well, that's what wound up on the cover of *Life* magazine," he would later reflect. "You can see the back of George Cox's head and Shepard coming. You can't really tell whether he's smiling or not, but he was almost in the helicopter and was pretty happy about that." [26]

Perched precariously on the sill of the capsule's hatch, Shepard had waited a few moments before grabbing the "horse collar," which dunked in the water before being lifted clear. "I grabbed it and got into it with very little difficulty. Shortly thereafter I was lifted directly from a sitting position out of the capsule up toward the chopper. The only thing that gave me any problem at all, and it was only a minor one, was that I banged into the HF antenna, but of course it is so flexible it didn't give me any trouble." [27]

Koons later reflected that the calm sea was ideal for the recovery, which went just as in training. "Within two minutes of the time [*Freedom 7*] hit the water we had the commander out of his capsule and in our craft." [28]

As Dean Conger recalls, the remarkable photographs of Shepard's retrieval which would grace many magazine covers almost didn't happen. "The antenna was broken off either before or after [ocean] impact. So the prepared plan was ditched and in the excitement of the event the copilot forgot about the switch for the camera and they began all the other recovery procedures. Fortunately, he remembered at the very last minute.

There were about 10 frames of Shepard coming up, and 230 frames of just the capsule and water after he was in the chopper! Doesn't matter. It was enough. Marine Lt. George Cox should get much of the credit for the success of the photos. The same bracket was used on subsequent flights but never produced a publishable photograph." [29]

In the top photograph, Shepard can be seen ascending to the hovering helicopter. At bottom, with the astronaut safely on board, the helicopter hoists *Freedom 7* from the sea and water streams out of the deployed landing bag. (Photos: U.S. Navy)

Once on board the helicopter, Shepard shook Cox's hand before being directed to a bucket seat. All members of the recovery crews had been given strict instructions not to speak to the astronaut unless he spoke to them first. Understandably, doctors and psychologists desired him to tell his story to them without it being colored by impressions conveyed to him during his return. "We were instructed not to direct our conversation to him," Cox explains, "but if he spoke to us we could answer him and talk to him if he started it. I pointed him toward his seat, to sit down for the ride back to the carrier." [30] Before he took his seat, however, Shepard looked out and said with consummate gratitude, "What a beautiful day!" Meanwhile, Cox and Koons attended to retrieving *Freedom 7* from the water.

As Shepard later recorded, "I sank into a bucket seat as soon as I reached the top, and on the way to the carrier I felt relieved and happy. I knew I'd done a pretty good job. The Mercury flight systems had worked out even better than we'd thought they would. And we'd put on a good demonstration of our capability right out in the open where the whole world could watch us taking our chances." [31]

WELCOME ABOARD, COMMANDER

George Cox remained on the Sikorsky's lower deck with Shepard for the short flight back to the carrier. Within seven minutes of retrieving the spacecraft from the water, the helicopter was zeroing in on the USS *Lake Champlain*.

As Shepard later pointed out, "When we approached the ship, I could see sailors crowding the deck, applauding and cheering and waving their caps. I felt a real lump in my throat." [32] He waved to the men as the pilots prepared to land.

The ship's crew watched the entire recovery process with great excitement. (Photo courtesy of Larry Kreitzberg)

On board "The Champ," there was outstanding reason to celebrate, as related by Scott Thompson from Beaver Falls, Pennsylvania. First, there had been the sight of *Freedom 7*'s splashdown. "We knew about where to watch. We saw this little speck coming down from the sky. Then we saw the parachute open and float down. When the capsule hit the water, there was a lot of steam because it was so hot." Soon after, an announcement blared out over the ship's loudspeakers reporting that Shepard was okay, giving rise to loud cheers. "Everybody went crazy because they were so happy. They knew it was an historic event – the first U.S. man in space. They could have heard us a long way off. We made a *lot* of noise." [33]

As Wayne Koons prepares to lower *Freedom 7* onto the waiting platform, a Navy helicopter shadows the Marine helicopter, taking photographs. (Photo courtesy of Ed Killian)

As in rehearsals, Koons carefully lowered *Freedom 7* onto the specially prepared platform that had been cushioned with mattresses. On Cox's command, he released the carrying hook. "The cargo hook could only be released by the pilot," Koons says. "The basic hook design, installed on all Marine HUS helicopters, had two methods of release. The first method was electrical, actuated by a button on the pilot's cyclic stick. The second method of release was mechanical, actuated by a foot pedal on the floor near the pilot's right heel. For Mercury retrievals, the electrical circuitry was disconnected. A special latch was installed on the cargo hooks for Mercury retrieval work. This latch prevented opening the hook until it was released by a lever on the cyclic. After releasing the latch, the hook could be opened by the foot pedal." [34]

Once the spacecraft had been safely situated on the platform and released, Koons set the helicopter down on the deck in front of 1,200 raucous sailors.

Freedom 7 had landed four nautical miles from where the USS *Lake Champlain* was stationed. The recorded departure time for the Marine helicopters that flew out to retrieve Shepard was 0927 (according to the OPNAV Part C record). Splashdown was at 0949, and the helicopter's arrival back at the carrier was at 1000, a total of 33 minutes.

Koons has a particularly fond memory. "I was busy shutting the helicopter down and here Shepard in his silver suit minus the hard hat comes slithering up ... through the space where George would have been if he were going to get up in his seat. He reached over and whacked me on the leg and [said], 'Good boy.' Then back down he went." [35]

Another person on board that day, NASA representative Charles Tynan, also has a serendipitous recollection. Film cameras were able to document the flight of the helicopter which carried Shepard and *Freedom 7* to a safe touchdown on the carrier, but "the Movietone News photographer later sent his movie film off the ship in a COD [cargo] aircraft and talked about possibly winning a Pulitzer Prize. I heard that shipboard personnel put a suicide watch on him when he found out his camera had malfunctioned and his film canister contained blank film." [36] It was later rumored that the hapless fellow had filmed the entire operation but had failed to remove his lens cover.

NASA, meanwhile, had issued an updated press bulletin declaring, "Test No. 108 is terminated. This was the pioneer U.S. man-in-space flight. The Mercury spacecraft is on the deck of the aircraft carrier *Lake Champlain* and the helicopter is about to land. Shepard is about to come out of helicopter." [37]

As the chopper was powered down, Navy physician Robert C. Laning and Army physician M. Jerome Strong approached and stood by the closed door. There was a moment or two of suspense before the door was suddenly flung open. George Cox climbed out, ready to assist the astronaut out of the helicopter, but he needn't have bothered; Shepard jauntily leapt down to the deck. Only eleven minutes had elapsed since splashing into the ocean. Standing on the deck, Shepard shook hands with Cox and gave him a heartfelt, "Thank-you, very much."

The magnitude of the welcome finally hit Shepard. "Until the moment I stepped out onto the flight deck of the carrier, I hadn't realized the intensity of the emotions and feelings that so many people had for me, for the other astronauts, and the whole manned

As Shepard shakes hands with George Cox, the medical staff moves in to escort the astronaut to the admiral's in-port cabin. (Photo courtesy of Dean Conger/NASA)

space program. This was the first sense I had of public response, of a public expression of thanks for what we were doing. I was very close to tears." [38]

According to the tight schedule, it was time for a medical checkup and to record a free-dictation report whilst the flight was still fresh in his mind. The two physicians approached, eager to escort Shepard to the admiral's cabin, but there was one last distraction for America's first astronaut. "I started for the quarters where the doctors would give me a quick once-over before I flew on to Grand Bahama Island for a full debriefing. But first I went back to the capsule, which had been gently lowered onto a pile of mattresses on the carrier's deck. I wanted to retrieve the helmet that I'd left in the cockpit. And I wanted to take one more look at *Freedom 7*. I was pretty proud of the job that *it* had done too." [39]

As the Navy helicopter hovers nearby, and much to the surprise of everyone present, Shepard returned briefly to *Freedom 7* to fetch his helmet from the capsule. (Photos courtesy of Ed Killian)

Shepard, helmet in hand, is escorted below deck by Dr. Jerome Strong (partially obscured). (Photo courtesy of Dean Conger/NASA)

Dean Conger (right, with camera) records Shepard departing the flight deck. (Photo courtesy of Howard Skidmore)

Ed Buckbee, who would go on to become the first director of the U.S. Space and Rocket Center and the founder of Space Camp, both located in Huntsville, Alabama, was a public affairs officer for the space agency at the time of Shepard's space shot. Some years later he asked the astronaut about climbing up and peering around the interior of his spacecraft. "Well, for one thing," Shepard responded, "a fighter pilot never leaves his helmet in the cockpit, so I reached in to get my helmet. I also looked around the instrument panel to see if I turned everything off." [40]

Shepard was then taken to the admiral's in-port cabin, located just forward of the port side deck-edge elevator and accessed by a catwalk running along the edge of the flight deck. It was here that he would disrobe and have his biomedical leads removed prior to medical checks. "I don't think you're going to have much to do," he told Dr. Laning with a wide grin as he consumed a refreshing glass of orange juice.

The priority task was to determine Shepard's condition immediately after having undergone high acceleration forces at launch, weightlessness, and deceleration loads. It had been feared that even a few minutes of weightlessness might possibly cause a lingering disorientation and perhaps even affect an astronaut's mind. But Shepard reported that he hardly realized when he had begun experiencing weightlessness, and his five minutes of zero-g proved to have left no trace of physiological or mental impairment. "It was painless," he pointed out. "Just a pleasant ride." He said the first real indication of being in a weightless state – he was tightly strapped in – was when a

As Shepard dictates his immediate reactions into a tape recorder, Dr. Laning and Dr. Strong assist him to remove his space suit and bio-med sensors. (Photos courtesy of Dean Conger/NASA)

stray washer floated by his left ear. In his opinion, having taken direct control of his spacecraft, an astronaut was fully capable of functioning freely in a weightless condition.

After this brief examination, the two doctors had to concur with Shepard's earlier remark that they wouldn't have much to do. Although he had arrived in the admiral's cabin perspiring and with a high pulse rate, that had soon settled once he was finally able to relax. He was in his usual superb physical condition. A more detailed medical examination was to be made when he arrived later that day at Grand Bahama Island. With the tests done, Shepard dictated his remaining impressions of the flight into a tape recorder.

Meanwhile, there were mixed feelings of pride, joy and relief for his wife at their ranch-style home in Virginia Beach. Once she had composed herself, Louise went out onto the front porch and the waiting crowd of news reporters and photographers swarmed in to capture her mood in their notebooks and cameras. "I don't have to tell you how I feel," she said, with a wide, happy smile on her face. "It's just wonderful. It's beautiful ... just wonderful." [41]

An excited Louise Shepard on the porch of their Virginia Beach home after hearing of her husband's recovery. She is accompanied by her parents, Mr. and Mrs. Russell Brewer, her niece Alice, and daughter Juliana. (Photo: Associated Press)

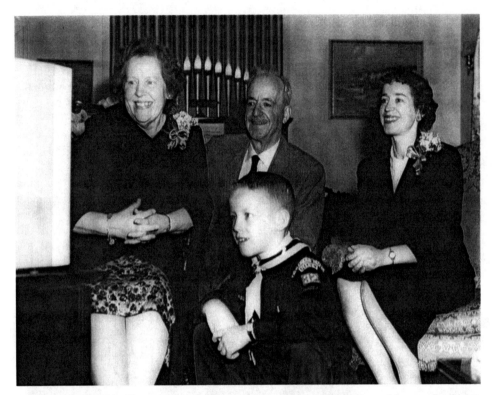

In East Derry, New Hampshire, Alan Shepard's parents, his sister Polly, and her son David, 10, are all smiles after news of his safe return to Earth. (Photo: United Press International)

In Shepard's hometown of East Derry, New Hampshire, the whole town exploded into a full-scale holiday. The streets were filled with rapturous people whooping and cheering and shaking hands with everyone they met, whilst church bells pealed out their glad tidings, fire engines wailed, and car horns added to the cacophony. There were calls for the town to be renamed "Spacetown, U.S.A."

SECURING THE SPACECRAFT

Meanwhile, on the flight deck, *Freedom 7* was being fully secured on its platform by the ship's special work detail personnel. As recalled by Ed Killian, "NASA technical representatives began to examine the capsule and record the final settings of the switches and gauge readings on the control panel and consoles. Marines were posted at the capsule, and the ship's special work detail and flight deck directors stood by to assist. NASA, Dean Conger and Navy photographers converged on the capsule."

With Shepard below for his debriefing, flight deck crewmen worked to steady the spacecraft and make it more secure on the platform. (Photo courtesy of Ed Killian)

Charles Tynan, NASA's Recovery Team Leader, carefully entered *Freedom 7* in order to verify settings and the condition of equipment. "Access to the capsule was not severely restricted as long as the NASA personnel were not interfered with in the performance of their duties," notes Killian. "The ship's crew could get close enough to peer inside and to photograph the *Freedom 7* capsule. They gathered nearby as the NASA Tech Rep made his inspection."

After the other helicopters had landed in their marked positions on the forward flight deck, the platform and its spacecraft cargo were rolled inboard and parked next to the island structure. Once the choppers had left their spots, the deck was clear for the fixed-wing aircraft that were to take off later.

"Once the platform was secured near the island," Killian continued, "the NASA technician resumed his examination of the capsule. At the top were the two empty quadrants were the parachutes had been housed. We could also see the periscope that the astronaut used to view the Earth on his ascent and descent. A bucket was placed near the capsule and unexpended green dye marker was bled off into the bucket. The capsule was on four-by-four wooden beams in order to prevent the landing bag from being damaged by the weight of the capsule. We could get close enough for an inside view of the capsule and to take pictures of its instrument panel. We could see where Shepard had been seated in the capsule. His head rested evenly with the window in the rear of the capsule."

At the same time, flight deck personnel were preparing the COD plane that was to fly Shepard and his NASA entourage to Grand Bahama Island, also known as GBI.

After NASA representative Charles Tynan had finished his inspection of the interior of the *Freedom 7*, the ship's crew were permitted a close inspection. (Photo courtesy of Ed Killian)

CREWMEMBER MEMORIES

Marine PFC Paul Molnoski was on guard duty on the USS *Lake Champlain* during the entire recovery operation, and his account recalls the contingency procedures that people were to follow in the event that the astronaut was found dead or badly injured after splashdown. This was similar in some ways to the sorrowful speech President Richard Nixon would record in case the two Apollo 11 moonwalkers were stranded on the lunar surface.

As Molnoski explains, "I was assigned to the forward port-side of the flight deck, and was walking my post when I heard over the loudspeaker that the countdown was held at minus-15, just fifteen minutes before the scheduled time of the launching at 7.30 a.m. At about 9.30 it was reported that the rocket had been fired and, a minute later, that the Redstone Mercury missile was airborne. We were instructed that when the capsule hit, if he didn't get out, Shepard would be presumed to be either dead or wounded. [The capsule] would be picked up by a Marine helicopter, brought aboard ship and taken down elevator number three. It would be opened and [Shepard] taken to sick bay. Only three persons beside the admiral would be allowed to speak to him; two doctors and one corpsman appointed by Washington. It would have been my job to clear the flight deck.

"No one was to cheer or try to talk with him when he came aboard ship. We heard his voice when he started reporting from the capsule. The first thing we heard was, 'feel fine; nothing unusual has happened.' I felt better then [because] I felt he would make it safely.

"I first spotted the capsule when it was about 4,000 feet up, descending under its parachute. He was six miles dead ahead of the ship. Three Marine and two Navy helicopters saw him. He hit the water about five or ten minutes later, surrounded by the 'copters. They had their hooks down and were ready to take him. When he hit the water, we waited for him to get out of the capsule. When he climbed out [and was hoisted] into the helicopter everyone was cheering. We felt great that he had made it. No one applauded or said anything when he came aboard, but everyone seemed to be taking pictures. He went right to the admiral's cabin. I was just about ten feet from him. Our job continued until 5.00 a.m. the next morning. We stood guard over the capsule." [42]

Anthony Vitulli was also on the ship that day. A 22-year-old graduate of the New York Institute of Photography, he was one of seven Navy photographers selected to document Shepard's recovery and recalls the day with fondness. "I was standing on the 07 level [the seventh deck of the island] with a 4-by-5 camera and photographed the capsule as it came out of the sky. We could see it clearly – the landing was *that* accurate." In addition to photographing the recovery, Vitulli also took pictures of the interior of *Freedom 7* once it had been secured on deck. "It was cramped," he said. "You look at that [capsule] and then you look at the [Space Shuttle] *Enterprise* and you say, 'Oh! How can that be?' The instruments were crude. You just can't believe someone went up in space in something like that." [43]

Prior to being assigned to the USS *Lake Champlain*, Michael Richmond had Navy recruit training at the Great Lakes Training Center on the shores of Lake Michigan, Illinois. He was part of the arresting gear handling crew – the cables and hardware used to arrest and rapidly slow the forward motion of a landing aircraft. When they

Air Officer Cdr. Howard Skidmore (right) with the Marine recovery pilots Wayne Koons and George Cox. (Photo courtesy of Ed Killian)

knew they were going to recover the *Freedom 7* spacecraft he and his crewmates had their cameras at the ready. "We went on the flight deck when they started bringing her in," he recalls. They watched the sequence of events with interest until Shepard headed below for his health checkup, leaving his spacecraft behind. "That's when we started taking pictures. They really didn't make a big issue about security or guards. We just walked up to it and took pictures." Like many other sailors, Richmond had a picture taken of himself proudly standing beside the history-making spacecraft [44].

Larry Kreitzberg from New York was a Navy photographer PH3 on board ship the day Shepard made his epic space flight. Several of his fine photographs appear in this book. As he explains, "I was assigned for this historic event to the 07 level of 'The Champ' along with several other ship's photographers who were positioned in various parts of the ship as well as in two helicopters. With my aerial camera I was waiting anxiously to see Alan Shepard and his Mercury capsule, *Freedom 7*. This I consider to be the most exciting time I spent in the Navy."

As the helicopter bearing its precious cargo approached the ship, "I looked down on the flight deck and observed the crew pointing upwards and watching the capsule being [lowered] on a frame covered with mattresses for cushioning. You could hear yelling and screaming as everyone was overcome with joy. I know I had tears in my eyes (which I'm not ashamed to say) along with everyone else. Watching Shepard depart the helicopter in his silver flight suit, smiling and waving, was one hell of a proud moment for all. I can say with pride that I was there for the first historic U.S. manned space flight with Commander Alan Shepard at the controls. That day is part of me and my life which I will never forget. A proud sailor, I was." [45]

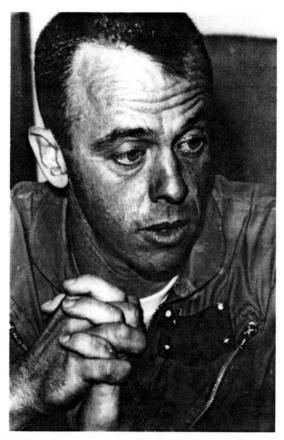

Wearing a clean flight suit, Shepard prepares for a call from the White House. (Photo courtesy of Dean Conger/NASA)

A PHONE CALL FROM THE PRESIDENT

Within an hour of arriving on board the carrier, Shepard received a radiotelephone call from President Kennedy in the White House, who wanted to extend his personal congratulations:

Kennedy: Hello, Commander.
Shepard: Yes, sir.
Kennedy: I want to congratulate you very much.
Shepard: Thank you very much, Mr. President.
Kennedy: We watched you on TV of course [at launch] and we are awfully pleased and proud of what you did.
Shepard: Well, thank-you, sir. As you know by now, everything worked just about perfectly and it was a very rewarding experience for me and the people who made it possible.

Kennedy: We are looking forward to seeing you up here, Commander.
Shepard: Thank-you very much. I am looking forward to it, I assure you.
Kennedy: The members of the National Security Council are meeting on another matter this morning and they all want me to give you their congratulations.
Shepard: Thank-you very much, sir, and I'm looking forward to meeting you in the near future.
Kennedy: Thank-you, Commander, and good luck.

Alan Shepard talking to President Kennedy. (Photo courtesy of Dean Conger/NASA)

At a press conference earlier that day, about 20 minutes after the safe recovery of Alan Shepard, the president issued a statement:

"All America rejoices in this successful flight of astronaut Shepard. This is an historic milestone in our own exploration into space. But America still needs to work with the utmost speed and vigor in the further development of our space program. Today's flight should provide incentive to everyone in our nation concerned with this program to redouble their efforts in this vital field. Important scientific material has been obtained during this flight and this will be made available to the world's scientific community.

"We extend special congratulations to astronaut Shepard and best wishes to his family who lived through this most difficult time with him. Our thanks also go to the other astronauts who worked so hard as a team in this project." [46]

The president said at his press conference that a substantially larger effort would be made in the space program, and that Congress would be asked for an additional appropriation. The request at that time was for 1.23 billion dollars, but he had earlier been advised that it would cost between 20 and 40 billion dollars to put a man on the Moon.

Whilst it should have been a joyous day for the president, with the first American astronaut safely back from space, he seemed to be anything but joyous at his press conference. He assured his audience that clearly he was happy about what had taken place, but he cited the tremendous courage and the accomplishment of Yuri Gagarin, and said that the United States was still a long way behind. Nevertheless, Shepard's flight was just the tonic that America – and its leader – needed in order to overcome earlier disappointments and the frustrations that had been mounting since the Soviet orbiting of Gagarin and the Cuban Bay of Pigs fiasco the previous month, which was still causing immense grief and humiliation for the young, newly installed president.

Alexander Wiley of the Senate Aeronautical and Space Science Committee, was openly ecstatic, declaring, "I'm on a sort of emotional drunk." Senator Karl Mundt, added that Alan Shepard ought to be awarded the Congressional Medal of Honor, saying that there were not enough words of praise for the bravery of the astronaut [47]. This, however, would have required a special act of Congress, so it was later decided to award him instead with NASA's Distinguished Service Medal.[2]

MOVING RIGHT ALONG

Shepard, meanwhile, was preparing to depart the USS *Lake Champlain* for Grand Bahama Island, where he would undergo a far more intense medical examination and debriefing.

Most of the NASA personnel were escorted to the COD, a Grumman TF1 Trader, by a beaming R/Adm George Koch, with Capt. Weymouth in attendance. Then the

---

[2] Shepard and his fellow astronauts were later awarded the Congressional Space Medal of Honor that was authorized by the U.S. Congress in 1969. Shepard received his from President Jimmy Carter in 1978.

astronaut made his appearance, climbing up the deck-edge ladder from the admiral's cabin. The crew saw he had discarded his silver space suit, although he still wore his silver flight boots, and was now wearing a more comfortable orange flight suit with a leather patch on the left breast which said ALAN SHEPARD, ASTRONAUT, USA. The flight suit and the name tag had been somewhat hastily made aboard ship. Since Shepard had urinated in his space suit during the extended pre-launch delay, it was a welcome change of clothing for him.

Still curious about the condition of his spacecraft, Shepard made a diversion over to where *Freedom 7* rested on its platform and shook hands with Charles Tynan, as Admiral Koch and Capt. Weymouth looked on. The astronaut and technician spoke briefly, then both men mounted the platform for a closer inspection of the spacecraft.

As Tynan informed the author, everything was not exactly as it may have seemed. "Since Shepard was the first U.S. astronaut we were told not to speak to him because they did not want anyone to 'cloud his mind' with thoughts other than the flight. I had finished getting in the capsule to record all switch positions and gauge readings, and I was standing near the capsule when Shepard appeared. He had finished with the doctors and wanted to examine the capsule before leaving the ship. He started a long conversation with me, telling me how wonderful the 'too short' flight was and that he was pleased that he came down within sight of the carrier. He said he rejoiced when the main parachute deployed."

Capt. Weymouth shakes the hands of Richard Mittauer from NASA's Public Affairs Office as Admiral Koch (rear, center) looks on. (Photo courtesy of Ed Killian)

Shepard talking to NASA Recovery Team Leader Charles Tynan. Capt. Weymouth stands at the rear. (Photo courtesy of Ed Killian)

Tynan said a later TV report on the flight mentioned that he and Shepard seemed to have been engaged in a very technical discussion. "Far from the truth: he wanted me to take the 8-day clock out of the capsule, bring it back to the Cape and give it to him. I was hesitant to start taking parts out of the first space flight capsule, resisting his request, and *that* is why our discussion lasted so long. Because the clock had no significant value to the flight, I removed a few screws and it was in my briefcase in no time. I gave it to Shepard a few days later in Hangar S. I understand that the seven astronauts had the clock mounted in a piece of walnut for their attorney's desk, who wasn't charging them for his work. I believe his name was something like [D'Orsay] who owned a small interest in the Washington Redskins football team." [48]

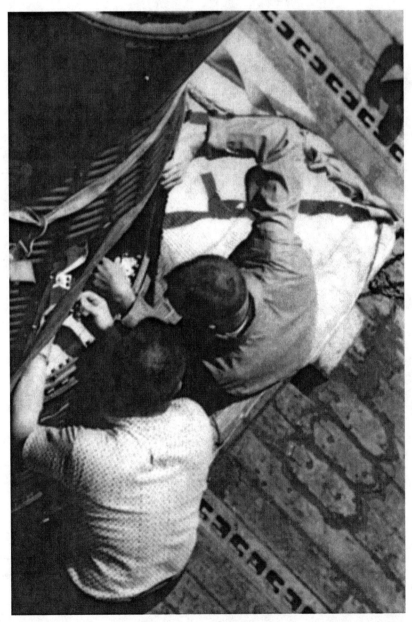

Shepard and Tynan peer into the spacecraft's interior. (Photo courtesy of Ed Killian)

Soon after, Ed Killian gained a prized memento of the day as Dean Conger took photographs from below of Shepard standing on the platform. "At the same time, those of us in Pri-Fly were snapping photos from above. Conger's photo of Shepard leaning into the capsule shows Pri-Fly just above the capsule. We moved back along the 05 level catwalk on the island and snapped a couple of pictures of Shepard at the capsule. Conger snapped a picture as he turned away from the capsule; this picture captured the Assistant Air Boss Tom Cooper, Air Controlman Russ Duncan and Air Controlman Ed Killian taking pictures from the catwalk just outside Pri-Fly and above Shepard."

Camera-bearing crew members can be seen on the catwalk taking their photographs. (Photo courtesy of Dean Conger/NASA)

This photograph, taken from the catwalk, shows a happy Alan Shepard after he had concluded his "business" with Charles Tynan. (Photo courtesy of Ed Killian)

Before leaving, Shepard found time to have a brief conversation with the ship's commanding officer, Capt. Weymouth. "He told me that four or five years from now we may look back on this as a pretty crude thing," Weymouth later revealed, "but at this moment it seemed a tremendous event." [49]

The ship's Executive Officer, Cdr. Landis E. Doner, R/Adm Koch, and his Chief of Staff then presented Shepard with records of his flight and its recovery.

Finally, having completed his inspection of *Freedom 7*, Shepard walked aft to the airplane that would take him to GBI for a battery of medical tests to be carried out in a special one-man "hospital". Two of the three HMR(L)-262 Marine helicopters and one S2F Tracker from VS-32 were to accompany Shepard's TF1 to Grand Bahama Auxiliary Air Force Base, about 75 miles southwest of the recovery site.

As Shepard's aircraft made a rolling takeoff from the USS *Lake Champlain* to the cheers of her crew, he had spent a mere 2 hours 25 minutes aboard the carrier. About 40 minutes later, another S2F from VS-22 flew off to deliver photographic film to Patrick Air Force Base at Cape Canaveral.

SHEPARD'S SECOND JOURNEY

In 1970 Cdr. Ted Wilbur reflected on transporting Alan Shepard on his aircraft from the USS *Lake Champlain* to Grand Bahama Island, "No sooner had I cleared the bow than he was out of his seat in the cabin and up to the cockpit, with that big wide grin spread

The TF1 Trader COD carrying Shepard is in the lead, preparing for a rolling takeoff. (Photo courtesy of Ed Killian)

across his face. Shouting above the noise of the COD's engines, he described his morning's monumental adventure, and it was easy to see he had been on top of the world, literally.

"*National Geographic* photographer Dean Conger was on board too, and after a series of pictures were taken, I pointed up ahead to where the Bahamas were coming into view. By then it was mid-afternoon and, as usual, tall [cloud] build-ups were forming over each island. I commented to Shepard that it would be a shame to spoil his day by running into a batch of bad weather. (The strip at Grand Bahama has no instrument facility.) He looked the situation over thoughtfully, then laughed: 'Swell! Let's divert to Nassau and pitch a liberty!' Unfortunately, we made it into GBI in good shape." [50]

Dean Conger was pictured shaking Shepard's hand on the aircraft. "Alan and I chatted away," he recalled. "But I don't remember any of what we said." [51]

THE SKIPPER AND A TIGHT SQUEEZE

Back on the USS *Lake Champlain* the spacecraft had been transferred below to the hangar bay, where Capt. Weymouth and his Executive Officer Cdr. Doner conferred briefly with Charles Tynan, NASA's senior representative. Weymouth then climbed into *Freedom 7* to get a feel for the cockpit, which he found was a little too small for him. He exited the hatch with the assistance of Cdr. Doner, who then took his turn to squeeze himself into the spacecraft.

A beaming Alan Shepard jokes with fellow passengers on the way to GBI. (Photo: Dean Conger/NASA)

Astronaut meets photographer: Alan Shepard shakes hands with Dean Conger. (Photo courtesy of Dean Conger/NASA)

After the ship had closed to within about eight miles of Florida during the night, Wayne Koons lifted off in Marine Corps helicopter #44 and as he hovered directly over *Freedom 7* George Cox attached the harness to the spacecraft so that it could be ferried back to Cape Canaveral.

"The helicopter had to get very close to the capsule to connect the harness with the shepherd's hook," Ed Killian recalled. "Although that day one should really have renamed that apparatus a '*Shepard*'s hook.' Anyway, the helo hovered, while the tension on the sling was taken up. The helo then moved to starboard over the capsule and lifted it clear of the platform. Helo and capsule were vectored toward the beach, accompanied by an escort. The platform that had held the capsule was then moved below as the helo and its package swung off into the distance. The capsule would be displayed for a period of time at Cape Canaveral."

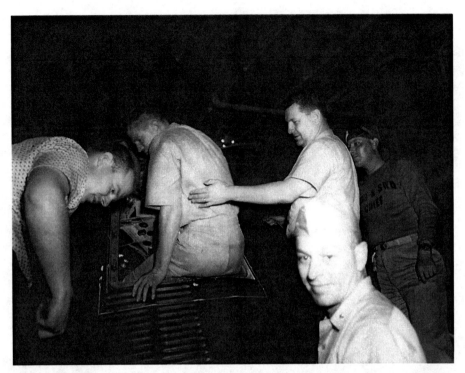

Capt. Weymouth exits the spacecraft with the assistance of Cdr. Landis Doner while Charles Tynan busies himself at left. (Photo courtesy of Ed Killian)

Pilot Wayne Koons eases *Freedom 7* off the landing pad ahead of the delivery flight across to Cape Canaveral. (Photo courtesy of Ed Killian)

As Wayne Koons points out, "I don't remember exactly where we set it down, but that's when the news coverage came in earnest. There were print reporters and TV crews. At that time they did everything on sixteen-millimeter cameras. So they'd get us out with these cameras, and we did lots of interviews. It was a heady time." [52]

Despite *Freedom 7* being the most celebrated piece of hardware in the world that day, its arrival at the Cape passed almost unheralded. Millions of television viewers had watched it ride atop the Redstone booster carrying Alan Shepard into space, but only 35 people – newsmen, engineers, guards – were on hand for its return. Set down just two miles from the launch pad that it had left the previous morning, *Freedom 7* was given a brief examination and then hauled off to a hangar where, in the weeks to come, the engineers and technicians would go over it inch by inch.

For Ed Killian and the USS *Lake Champlain*, things quickly returned to normal. "That ended our participation in the first U.S. manned space flight and we headed home. It had been a long cruise, but we'd been a part of something truly historic."

A recent photo of Ed and Kath Killian. (Photo courtesy of Ed Killian)

Frank Yaquiant of Baltimore, Maryland agrees. He joined the carrier in May 1960 as a member of the V-4 division, responsible for the aviation fuel the planes used. "I was very happy to have been on 'The Champ' that day. My shipmates and I were witnesses to something truly special. It's been almost 52 years since that memorable day, and the science of space travel has advanced far beyond that available during the flight of *Freedom 7*. The achievements of those past 50-plus years may make the events of May 5, 1961 seem rather modest to the uninformed [today]. But this was America's first venture into manned space travel, and those of us aboard the USS *Lake Champlain* that day have the pride and satisfaction of knowing we were there at the beginning." [53]

As a footnote to the story, about ten years after Shepard's flight, Ed Killian was dining with his family at Vargo's restaurant in Houston when he became aware that Alan Shepard had just walked in. As Shepard was talking to a party of people who had greeted him at the door, Killian decided to introduce himself and strolled over. "As I drew near, the crowd began to break up. I offered my hand and said, 'Admiral, I'm Ed Killian, and we haven't met, but you may remember '*Mercury, Mercury*, this is *Nighthawk*. Do you read?' He smiled and said 'How could I forget?' Then his eyes narrowed as if remembering more of the details. 'That was you?' I nodded, smiling. 'Hmmm,' he said, 'I was pretty excited, wasn't I?' He tilted his head and raised his eyebrows, as if to ask a further question. I sensed that he was making an oblique reference to the private conversation we had had. 'You had a right to be excited. So did we all,' I observed. Satisfied with my response, he said, 'Well, very nice to meet you,' nodding and shaking my hand. He then turned back to his party and they were ushered to his table. I knew there was a wild world of differences between us, but I also knew the two of us shared a secret."

References

1. Killian, Ed, "Mercury Three Recovery", 2010, in website *http://tech.groups.yahoo.com/group/Freedom_7_Recovery/files*. Extracts from his 1961 unpublished manuscript and photo collection throughout chapter reprinted with permission
2. Swenson, Loyd S., James M. Grimwood and Charles C. Alexander, *This New Ocean: A History of Project Mercury*, NASA History Office, Washington, D.C., 1998
3. Tanner, Beccy, article, "Lyons man flew helicopter that picked up astronaut Alan Shepard," *The Wichita Eagle* newspaper, Kansas, 6 May 2011
4. Kilbourn, Clara, article, "Ottawa University honors NASA astronaut", *The Hutchinson News* newspaper, Ottawa, Kansas, 2 May 2009
5. E-mail correspondence with Wayne Koons, 5 February 2013
6. Koons, Wayne, interviewed by Rebecca Wright for NASA JSC Oral History program, Houston, Texas, 14 October 2004
7. E-mail correspondence with Wayne Koons, 5 February 2013
8. Koons, Wayne, interviewed by Rebecca Wright for NASA JSC Oral History program, Houston, Texas, 14 October 2004
9. E-mail correspondence with Dean Conger, 5 February 2013
10. Powers, Scott, article, "NASA's glory decade began with Alan Shepard's launch 50 years ago", *Orlando Sun* newspaper, Florida, 5 May 2011

11. Shepard, Alan, "First Step to the Moon", *American Heritage Magazine*, Vol. 45, Issue 4, July/August 1994
12. *Ibid*
13. Shepard, Alan B., USS *Lake Champlain* debriefing, 5 May 1961. From *Exploring the Unknown: Selected Documents in the History of the Civil U.S. Space Program*, edited by John M. Logsdon, NASA History Series, SP-4407, Washington, D.C., 1999
14. Carpenter, M., L. Cooper, J. Glenn, V. Grissom, W. Schirra, A. Shepard, D. Slayton, *We Seven*, Simon and Schuster, New York, NY, 1962
15. *Ibid*
16. CBS network television program, *I've Got a Secret*, recorded New York, NY, 10 May 1961
17. Koons, Wayne, interviewed by Rebecca Wright for NASA JSC Oral History program, Houston, Texas, 14 October 2004
18. NASA Project Mercury Working Paper No. 192, *Postflight Report for Mercury-Redstone No. 3 (MR-3)*, NASA Space Task Group, Langley Field, Virginia, 16 June 1961; declassified 19 June 1973.
19. Koons, Wayne, interviewed by Rebecca White for NASA Oral History program, Houston, Texas, 14 October 2004
20. Email correspondence with Charles Tynan, 21 December 2012
21. CBS network television program, *I've Got a Secret*, recorded New York, NY, 10 May 1961
22. Shepard, Alan B., USS *Lake Champlain* debriefing, 5 May 1961. From *Exploring the Unknown: Selected Documents in the History of the Civil U.S. Space Program*, edited by John M. Logsdon, NASA History Series, SP-4407, Washington, D.C., 1999
23. E-mail correspondence with Wayne Koons, 5 February 2013
24. Shepard, Alan B., USS *Lake Champlain* debriefing, 5 May 1961. From *Exploring the Unknown: Selected Documents in the History of the Civil U.S. Space Program*, edited by John M. Logsdon, NASA History Series, SP-4407, Washington, D.C., 1999
25. CBS network television program, *I've Got a Secret*, recorded New York, NY, 10 May 1961
26. Koons, Wayne, interviewed by Rebecca Wright for NASA JSC Oral History program, Houston, Texas, 14 October 2004
27. Shepard, Alan B., USS *Lake Champlain* debriefing, 5 May 1961. From *Exploring the Unknown: Selected Documents in the History of the Civil U.S. Space Program*, edited by John M. Logsdon, NASA History Series, SP-4407, Washington, D.C., 1999
28. CBS network television program, *I've Got a Secret*, recorded New York, NY, 10 May 1961
29. E-mail correspondence with Dean Conger, 5 February 2013
30. CBS network television program, *I've Got a Secret*, recorded New York, NY, 10 May 1961
31. Carpenter, M., L. Cooper, J. Glenn, V. Grissom, W. Schirra, A. Shepard, D. Slayton, *We Seven*, Simon and Schuster, New York, NY, 1962
32. Carpenter, M., L. Cooper, J. Glenn, V. Grissom, W. Schirra, A. Shepard, D. Slayton, *We Seven*, Simon and Schuster, New York, NY, 1962

33. Utterback, Debra, article, "He remembers it well", *Beaver County Times* newspaper, Pennsylvania, 5 May 1986
34. E-mail correspondence with Wayne Koons, 5 February 2013
35. Koons, Wayne, interviewed by Rebecca Wright for NASA JSC Oral History program, Houston, Texas, 14 October 2004
36. Email correspondence with Charles Tynan, 23 December 2012
37. *Saskatoon Star-Phoenix* (Canada) unaccredited newspaper article, "U.S. Officer Rockets Briefly Across the Threshold of Space", 5 May 1961
38. Shepard, Alan, "First Step to the Moon", *American Heritage Magazine*, Vol. 45, issue 4, July/August 1994
39. *Ibid*
40. Buckbee, Ed with Wally Schirra, *The REAL Space Cowboys*, Apogee Books, Ontario, Canada, 2005
41. The *Telegraph-Herald* newspaper, unaccredited article, "Shepard Passes All Tests With Flying Colors – Doctors," Dubuque, Iowa, 7 May 1961
42. *The Hastings News* newspaper (Hastings, New York), unaccredited article, "Hastings Marine Saw Historic Space Flight", 22 July 1961, pg. 4
43. Sastrowardoyo, Hart, interview with Anthony Vitulli, Stafford, Virginia. Extracts used with permission
44. Smith, Steven, article for *Space Times News* online, 5 May 2011. Website: *http://www.space-timesnews.com/news/2011/may/05/florida-man-had-front-row-seat-1st-us-manned-space*
45. E-mail correspondence with Larry Kreitzberg, 5 December 2012
46. The London *Times* newspaper (U.K.), article "Mr. Kennedy telephones congratulations", 6 May 1961
47. *Ibid*
48. E-mail correspondence with Charles Tynan, 23 December 2012
49. Caidin, Martin, *Man Into Space*, Pyramid Books, NY, 1961, pg. 36
50. Wilbur, Ted, *Naval Aviation News* article, "Once a Fighter Pilot", Washington, DC, issue November 1970
51. E-mail correspondence with Dean Conger, 5 February 2013
52. Koons, Wayne, interviewed by Rebecca Wright for NASA JSC Oral History program, Houston, Texas, 14 October 2004
53. Correspondence received by author from Frank D. Yaquiant, Baltimore, MD, 2 January 2013

7

A nation celebrates

Alan Shepard's colleague Gus Grissom had monitored the liftoff of *Freedom 7* from inside the Mercury Control Center at Cape Canaveral. As prearranged, once Grissom knew that the mission was underway he left the building to make a short flight across to Grand Bahama Island. Once there, and while everyone waited for Shepard to also reach the island, Grissom was asked by reporters to comment on how he felt seeing his buddy launched into space, and when he thought his own chance might come. "I'm very happy," he said in reply. "You can underline that. I wanted to be the one chosen for this shot and I certainly want to be chosen the next time. Everything went perfectly, just like we practiced it a thousand times." [1]

GRAND BAHAMA ISLAND

Deke Slayton had also flown in ahead of Shepard's arrival, and he waited patiently alongside Grissom as the TF1 taxied in at the auxiliary air base. The two men were in high spirits after a day of high drama. Once Shepard had stepped down from the airplane they clapped him on the shoulder and knuckle-rubbed his crew-cut hair. For his part, Shepard playfully punched each man in the chest. The three grinned broadly and skipped around like small boys unable to contain their excitement. Slayton was heard to say that the flight had been "perfect – it couldn't have been any better. You pulled it off real good."

"Everything went fine," Shepard replied within earshot of reporters. He waved, but no interviews were permitted.

A host of technical and medical personnel were also present to greet the astronaut, including Bill Douglas and nurse Dee O'Hara, who has told the author that her joy in knowing that Shepard splashed down safely was "overwhelming … I was so relieved to see him." [2] Still joking and laughing, Slayton and Grissom accompanied Shepard to an Air Force station wagon belonging to Capt. Hugh May, the commander of the island missile tracking station. When they finally pulled up at the aluminum portable

Deke Slayton and Gus Grissom welcome Shepard to Grand Bahama Island. (Photo courtesy of Dean Conger/NASA)

medical facility where Shepard was to undergo an extensive medical and psychiatric evaluation he was ushered inside, despite reporters shouting questions in an effort to get him to say a few precious words. Within the hour, a brief but heavy rainstorm had beat a tattoo on the white roof of the specially erected hospital.

Shepard called his wife Louise by radio telephone after he had settled in on the island and they spoke for a while, but the connection was bad and only a few phrases came through clearly. He told her that he was pleased with the way the flight went and asked about their family. She told him that the family was just fine, they were all proud of him, and had watched the launch on television.

John Glenn had been asked to sleep in Shepard's hospital room that night, as the psychologists had suggested that as a precautionary measure the returned astronaut should not be in a room by himself. It was obviously an over-reaction on their part, but nobody really minded. This was all new stuff to everyone. However, the whole evening lay ahead of America's newest hero. When later asked for his feelings on the events of the day, Grissom told reporters, "I have to admit I'm a little jealous. I think I've a fair chance of being on the next launch. I do want to be on the next one. I wanted to be on this one." Slayton chimed in, "I wished I'd been up there, too." [3]

Grissom accompanies Shepard to the special medical facility. (Photo courtesy of Dean Conger/NASA)

Lt. Col. John ('Shorty') Powers, NASA's spokesmen for the astronauts, said he had never seen Shepard more cool and calm. He said Shepard's schedule for the first 24 hours would comprise of an extensive physical checkup lasting at least two full hours by Dr. Col. William Douglas, the astronauts' physician, and Dr. Maj. Carmault Jackson, an internist. Then Shepard would have a free half hour before a full hour of free dictation into a tape recorder recounting his experiences because, as Powers put it, "We want to make sure we don't lead him on his thoughts." Next, two engineers were to go over details with Shepard of the performance of the Redstone booster and of the Mercury spacecraft. On Saturday morning, following a good night's sleep, he would be interviewed by psychologists concerning his feelings and sensations [4].

Astronauts Carpenter, Cooper, Glenn and Schirra flew in later that day so that the entire group could hear Shepard give an account of his experiences and indicate what they might expect when making their own missions. Project Mercury engineers were also bringing data tapes recorded during his flight so that they could discuss different aspects of the mission from an engineering viewpoint.

That night the Air Force base personnel were invited to what was known tongue-in-cheek as the "Grand Bahama Yacht Club" to celebrate the successful space shot. It was only a bare-bones club with a bar, and none of the personnel actually owned a yacht, but there was a dartboard and card games to amuse everyone as they enjoyed their drinks. The members of the press gathered at the front gate of the base were not allowed in to take pictures or conduct interviews.

When Alan Shepard and his astronaut colleagues walked in through the front door of the club in the early evening, the general noise and laughter instantly turned into a standing ovation. Shepard joined Bill Douglas, Dee O'Hara, and other base officials at their table.

Dee O'Hara has a lingering memory of the occasion, "That evening, I remember, we were relaxing in an island bar with a little TV sitting on a plank in a corner. Alan, Bill Douglas and I were able to sit back and finally watch all the news of his flight. Alan was in such a good mood – it had been quite a day!

"It seemed so unreal, sitting there watching TV with Al, and the reporters were saying how the astronaut was now on Grand Bahama Island, and probably enjoying a glass of iced tea. And there we were, knocking down a quiet glass of Scotch and water! I used to smoke back then, and when I lit one up Al leaned over and said, 'Ah, could I have a puff of your cigarette, Dee?' He also smoked back in those days, and he hadn't had a cigarette since before his flight, so I said 'Sure!' He took one puff, had a swig of my Scotch and water, and I remember thinking to myself, 'Oh public – if you only knew!'" [5]

Dee O'Hara was correct; newspapers the following day reported that Shepard had dined that evening on a huge shrimp cocktail, a roast beef sandwich, and iced tea.

As arranged, John Glenn slept in the same room as Shepard that night. "Being backup meant you virtually lived with the person," Glenn later reflected. "While his flight had gone perfectly, uncertainties remained. The doctors preferred not to leave the astronaut completely on his own, even after the flight. Nobody knew what the delayed action might be.

"Al's reaction was exuberance and satisfaction. He talked about his five minutes of weightlessness as [being] painless and pleasant. He'd had no unusual sensations, was elated at being able to control the capsule's attitude, and was only sorry that the flight hadn't lasted longer." [6]

On Saturday morning, relaxing in a sports shirt and slacks, Shepard sat down for breakfast just before 8:30 a.m. and enjoyed a hearty meal of scrambled eggs, toast, jelly, and orange juice. Ahead of him lay a busy day filled with more medical checks and a host of interviews. In all, some thirty-two specialists would participate in the debriefing, including physicians, program managers, operations engineers, public relations personnel and official photographers. In addition to receiving a full medical from Bill Douglas and his team, he would be asked about his in-flight activities and performance, and about the performance of the vehicle systems.

Dr. George Ruff, a psychiatrist at the University of Pennsylvania, and Dr. Robert Voas, a NASA psychologist and training officer, were present to conduct extensive interviews with Shepard. His reflexes were also checked by Dr. Carmault Jackson, and his general well-being assessed by Dr. Phillip Cox. There would be another chest X-ray and more blood samples taken. Later that day, they agreed that the astronaut was in excellent health and good spirits. "He's just like he was before the flight, only he's happier, of course," reported Bill Douglas, who said the tape records of Shepard's flight "showed he performed remarkably well the complex tasks required of him. Five minutes of weightlessness apparently posed no problem, nor did the increased gravity pull of reentry." [7]

RESULTS OF POST-FLIGHT MEDICAL EXAMINATIONS

A comprehensive report on the pre-flight and post-flight medical condition of Alan Shepard was prepared by:

- Carmault B. Jackson, Jr., M.D. *Aerospace Medical Branch*
- William K. Douglas, M.D., *Astronaut Flight Surgeon*
- James F. Culver, M.D., *USAF Aerospace Medical Center, Brooks AFB, Texas*
- George Ruff, M.D., *University of Pennsylvania*
- Edward C. Knoblock, Ph.D., *Walter Reed Army Medical Center*
- Ashton Graybiel, M.D., *USN School of Aviation Medicine, Pensacola, Florida*

and in part, their report reads:

> The first post-flight physical examination was performed aboard the aircraft carrier *Lake Champlain*. Blood and urine specimens were collected and the pilot was asked to begin debriefing in the form of free dictation. Three hours from liftoff, Astronaut Shepard was taken to Grand Bahama Island by aircraft from the carrier. On arrival at this remote island site, he seemed quietly elated and offered no complaints. His own statement of general fitness included "a wonderful flight," "everything went well," "I feel fine."
>
> The psychiatrist at the time of his interview, which actually took place after the next physical examination, believed that the "subject felt calm and self-possessed. Some degree of excitement and exhilaration was noted. He was unusually cheerful and expressed delight that his performance during the flight had actually been better than he expected. It became apparent that he looked upon the flight as a difficult task about which he was confident, but could not be sure,

of success. He was more concerned about performing effectively than about external dangers. He reported moderate apprehension during the pre-flight period, which was consciously controlled by focusing his thoughts on technical details of his job. As a result, he felt very little anxiety during the immediate pre-flight period. After launch, he was preoccupied with his duties and felt concern only when he fell behind on one of his tasks. There were no unusual sensations regarding weightlessness, isolation, or separation from Earth. Again, no abnormalities of thought or impairment of intellectual functions were noted."

In physical terms, the physicians identified only minor fluctuations between the examinations given pre-flight, post-flight, and on Grand Bahama Island. One part of their reports reads:

> Mild dehydration and early signs of heat exhaustion were also evident when an individual in an impermeable Mercury pressure suit was not adequately ventilated. With Redstone training profiles, there has been no nystagmus as a result of high noise levels; there has been no vibration injury .... From the material obtained, it is obvious that a brief sortie has been made into a new environment. Similarities between this sortie and a previous training experience were noted. No conclusions have been drawn except that in this flight the pilot appears to have paid a very small physiologic price for his journey [8].

Alan Shepard relaxing on Grand Bahama Island. (Photo: United Press International)

REACTIONS ABROAD

In his first public comment on the space shot the day before, Shepard told newsmen, "The only complaint I have is that the flight wasn't long enough."

Grissom had a chance to relax with Shepard over a meal that day, and ask specific questions about the flight. He was next in line to fly, and his spacecraft was already undergoing final checks and tests for the MR-4 mission, slated for July. Two major changes had been made to his spacecraft, which he had already named *Liberty Bell 7*. In addition to its periscope, *Freedom 7* had had two small portholes. Instead, *Liberty Bell 7* had an enlarged, trapezoidal window to provide for better observations by the astronaut. And *Freedom 7*'s awkward, latch-operated hatch had been replaced by an explosive side hatch to enable the astronaut to make a rapid egress in the event that the capsule started to take on water. This was an innovative feature that all too soon would cause grievous and ongoing concerns for Grissom.

Meanwhile, reaction to Shepard's flight in Communist capitals tended towards admiration for the man, tepid praise for the feat itself, and smug comments stressing that his flight could not compete historically or technically with that of Yuri Gagarin the previous month. Soviet Premier Nikita Khrushchev indicated an awareness of the American space flight without mentioning it specifically. In lauding Gagarin and his orbital flight in a speech given at Erevan in Soviet Armenia, he noted Gagarin had flown "around the globe precisely – not just up and down."

The official Soviet news agency, TASS, reported, "The launching carried out in the United States of America [on] Friday of a rocket with a man aboard cannot be compared with the flight of the Soviet spaceship Vostok which carried the world's first cosmonaut, Yuri Gagarin. During Shepard's training for the flight, the American press itself acknowledged that from the point of view of technical complexity and scientific value, this flight would be very inferior to the flight of Gagarin.

"For example, *Time* magazine emphasized that the American project of sending a man into space was designed only to put a man into a short trajectory which is considerably less than the complicated flight of the Vostok round the Earth. To begin with, the Soviet ship orbited around the Earth at a maximum height of [203 miles]. The cosmonaut made a full orbit around our planet and only then landed in the pre-set area of the territory of the U.S.S.R. The rocket with the man on board which was launched in the United States of America is in reality like an intercontinental missile, since it covered a limited distance of the Earth's surface at a maximum height of 115 miles.

"The capsule which carried the American astronaut fell at a distance of only 302 miles from the site of the launching. The entire flight of the American rocket took only 15 minutes, while Yuri Gagarin spent 108 in his orbiting flight round the Earth.

"The following fact shows the principal difference between his flight and that of the American rocket: Yuri Gagarin was in a state of weightlessness during the entire time his spaceship was in orbit, but the American astronaut was in this state only several minutes." [9]

Not surprisingly, no mention was made of the fact that Gagarin's spacecraft was controlled from Earth throughout his flight, whereas Shepard had manipulated his spacecraft's roll, pitch and yaw movements using manual controls.

Soviet-sponsored broadcasts in Czechoslovakia also dismissed Shepard's shot as "primitive and outmoded," although Czech and Hungarian broadcasters recognized the man himself as a heroic figure. Meanwhile, the Red Chinese press in Hong Kong was disparaging. The *Communist Commercial Daily* described the flight as nothing more than a publicity stunt [10].

On Grand Bahama Island, all the tests showed that Alan Shepard was in perfect physical and mental health. Chris Kraft was NASA's first flight director, serving in Mission Control, and he later said that he was also pleased with Shepard's physical condition throughout the flight. "The only medical data that raised eyebrows was Shepard's heart rate, spiking at 220 and holding above 150 for several minutes. Both numbers were above anything seen before in Shepard's medical condition – even during stress tests and centrifuge runs. But the surgeons weren't that worried. They'd recently gotten a report on race car drivers showing heart rates even higher, and lasting for several hours. Even sucking in a bit of carbon monoxide didn't affect their high-speed driving. Based on that, Shepard's heart rate seemed well within the norm for a person whose adrenaline was flowing." [11]

TO WASHINGTON

After dinner on Saturday night, a relaxed Shepard joined others on wooden benches at the open-air movie lot at the base for a screening of *The Grass is Always Greener*, a comedy starring Cary Grant.

The next morning, once again casually dressed and indistinguishable from several hundred other men at the air base, Shepard had breakfast with Grissom in the mess hall. By tacit consent, everyone on the base had agreed to allow him to simply go his way normally without any bustle or crowding. It would also be his last day of leisure on the off-shore refuge.

On Monday, under orders from the President, Shepard flew to Washington, D.C., along with the other astronauts. At Andrews AFB, where they landed in suburban Maryland at 9:33 a.m., Shepard was the last one off the airplane. Waiting to meet him at the foot of the ramp was his wife Louise, her parents Phil and Julia Brewer, his parents Alan and Renza, and his sister, Polly Sherman. There were hugs, kisses and handshakes before they were ushered over to a microphone set up nearby so that he could briefly address a crowd of almost 1,000 cheering people who had gathered to greet him. He thanked everyone at the base for their welcome and delivered a few words of encouragement prior to escorting his wife to the waiting helicopters that would quickly convey them to the south lawn of the White House and a reception by the President and First Lady at which he would receive NASA's highest award, the Distinguished Service Medal. The only previous such medal had been awarded two years earlier to John W. Crowley for his 38-year contribution to the nation's aircraft, spacecraft and missile programs.

The presentation ceremony took place on a specially built platform in the Rose Garden, just outside the White House, and was conducted in a very relaxed manner. Kennedy and Shepard shook hands more like college classmates than President and hero. Shepard was relaxed and easy during the moments of greeting before it began. As he remarked, "I thought last Friday was a thrilling day, but this surpasses it." [12]

After the USS *Lake Champlain* had returned to Quonset Bay following the retrieval of Alan Shepard, his parents, Alan and Renza, were given a grand tour of the ship, which included a visit to the carrier's hangar bay where a mockup of *Freedom 7* was perched on the actual platform which had held their son's spacecraft after recovery. (Photo courtesy of Larry Kreitzberg)

216 A nation celebrates

Alan and Louise Shepard meet the Kennedys, with Vice President Lyndon Johnson looking on. In center of the picture is NASA's Public Affairs Officer, John 'Shorty' Powers. (Photo courtesy of Dean Conger/NASA)

With his fellow Mercury astronauts looking on, Shepard shakes hands with President Kennedy at the White House reception. (Photo courtesy of Dean Conger/NASA)

In his preamble, Kennedy said the nation was proud of Shepard and his fellow astronauts, noting that his flight was made under conditions of full publicity in a free society. The United States had "risked much and gained much," he stated. This was meant as an emphatic reminder that the Soviet Union, operating in a closed society, was following a tradition of secrecy in its space program that stood in spectacular contrast to the methods endorsed by the American nation. The President also paid a grateful tribute to the other astronauts and to the NASA officials who had brought the agency together and carried it through to that point. They were key members of the team that made the flight such a great success. To the 500 dignitaries gathered in the Rose Garden, the President said, "We should give them all a hand." And then, to distinguish the astronauts from the assembled bureaucrats, he jokingly described them as "the tanned and healthy ones – the others are Washington employees."

Kennedy then read the NASA citation, which pointed out that Shepard's flight "was an outstanding contribution to the advancement of human knowledge and space technology and a demonstration of man's capabilities in suborbital space flights." He then described the medal as "a civilian award for a great civilian accomplishment." [13]

At this point an air of informality took over, adding a special note of its own to the ceremony. During the presentation Kennedy fumbled for a moment, dropping the medal with its blue and white ribbon onto the wooden floor of the platform. But he quickly retrieved the medal and handed it to Shepard, saying he was proud to present "this decoration – which has gone from the ground up," thereby creating a storm of laughter. But then, instead of following custom and pinning the medal on Shepard's coat, Kennedy simply handed it over, saying, "Here." After Shepard had responded in acknowledgement of the honor bestowed on him, Jacqueline Kennedy whispered to her husband that he ought to have pinned the medal onto the astronaut's jacket. Taking it back from Shepard, the President quipped, "Let me pin it on; I'll do my duty." [14]

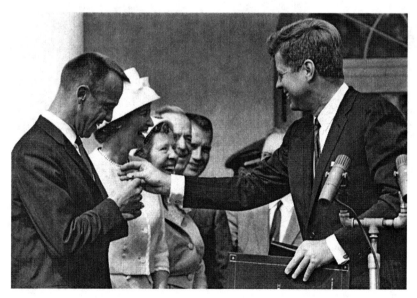

Amid laughter, the President presents Shepard with the NASA medal he had dropped on the ground. (Photo courtesy of Dean Conger/NASA)

A CAPITOL PRESS CONFERENCE

In 1986, Shepard reflected back on that auspicious day. "After the ceremony in the Rose Garden, the president invited the astronauts into the Oval Office to talk about the future of the space program. Kennedy and Vice President Lyndon Johnson had great political instincts. They knew the country needed a lift, and they saw space flight as a rallying point. We talked at great length about it. The president said he knew I had a parade up Pennsylvania Avenue, but first he wanted me to go with him to a meeting of the National Association of Broadcasters. He just grabbed me, and we got in his car and drove to the meeting. 'I want you to say a few words to these guys,' Kennedy said. I forget what I said; it was something like it was nice to be back. Everybody jumped to their feet and cheered. I couldn't believe the reception there." [15]

According to the 1994 book, *Moon Shot*, based on interviews with the astronaut, "Shepard did not like what was happening. His patience was evaporating swiftly. He disliked, intensely, being used. Walking in on the broadcasters' convention with the president would be showing off a war trophy named Shepard, and it smelled. He mollified himself somewhat by remembering that no matter who he was, Kennedy was also his commander in chief, and you can excuse almost anything if you're obeying orders. The fact that he and Louise received a standing ovation did diminish his objections to some degree." [16]

Vice President Lyndon Johnson sits between the Shepards as the motorcade prepares to leave the White House. (Photo: Associated Press)

Further cheers went up as Alan and Louise Shepard, with Vice President Johnson between them, climbed into a limousine which then left the White House for a slow ride along the traditional parade route of American heroes, from the White House out onto Pennsylvania and Constitution Avenues to the Capitol, where he was to receive a reception by Congress.

A massive, cheering crowd estimated at around a quarter of a million people lined the sidewalks. There were no military bands playing, no troops, and no spectacular displays of the nation's might. Government officials had earlier decided that it would be inappropriate to give their official blessing to a star-spangled military parade and organized welcome, so the citizens of Washington took matters into their own hands and by their tumultuous welcome gave the nation's first spacemen and his fellows in the following limousines a cacophony of unrehearsed affection that made the day's proceedings all the more memorable.

"I'll never forget riding to the Capitol in an open convertible with Johnson and Louise," Shepard later pointed out. "Johnson kept saying, 'Look at all these people ... Shepard, you and Louise get up on top of this thing.' So we sat up on the back. When we got to the Capitol, Johnson said, 'Well, Shepard, now that you're a famous man, let me give you some advice. Never pass up an opportunity for a free lunch or a chance to go to the men's room.'" [17]

The Shepards wave to crowds lining the motorcade route along Washington's 15th Street. (Photo: Associated Press)

A massive crowd had gathered at the foot of Capitol Building to catch a glimpse of the nation's first astronaut. (Photo: Associated Press)

On arrival at the east portico of the Capitol, Shepard alighted along with the other astronauts in front of a massive crowd. All seven would have been stunned by the thousands of excited people who had gathered in the plaza hoping to catch a glimpse of them but held back by police lines and ropes. Led by Vice President Johnson, the party ascended the broad steps of the Capitol to meet House Speaker Sam Rayburn, House Majority Leader John McCormack, and other dignitaries. Shepard expressed his sincere appreciation for the warm welcome by the lawmakers, and indicated he would have far more to say at his later press conference. He seemed quite composed in front of the noisy, jostling crowd, speaking seldom, smiling often, and watching the scene before him with amazement.

After waving at the crowds and saying a few words, Shepard moved on to the more formal part of the occasion, which was held in the old Supreme Court chamber, just a few steps down from the Senate. Then it was off to the State Department for his press conference, chaired and introduced by NASA Administrator James Webb. As Shepard entered the auditorium, 500 news reporters rose and gave him a standing ovation, something reporters seldom do. With his six fellow astronauts flanking him on his right and James Webb, Robert Gilruth and other NASA officials on his left he gave an account of the flight "we made" the previous Friday. He acknowledged the acclaim, but refused to accept it all for himself.

As Vice President Johnson looks on, Shepard prepares to give a news conference at the Capitol Building. (Photo: Associated Press)

To the President, to the Congress, and to the reporters, Shepard stressed over and over that it was not "he" but "we" who did the thing that they were praising him for. "I am acutely aware," he said, "of the hundreds of individuals who made this flight possible." But as Robert Gilruth, director of the Mercury program pointed out at the conference, it was Alan Shepard "who really broke the ice for all of us" and showed America the way into the great new frontier of space [18].

That night, after the glare of the public spotlight, Shepard and his wife and family spent a quiet evening together in the seclusion of Langley Air Force Base, Virginia, talking over recent events.

Even as the nation celebrated the successful first American space flight by Alan Shepard, most of the attention at NASA had turned to the future and was focused on the second Mercury flight and those which would follow it. Soon 40-year-old Virgil Grissom, better known as Gus, would step up to the plate to deliver the all-important second suborbital test in the spacecraft he had patriotically named *Liberty Bell 7*.

The NASA medal presented to Shepard earlier that day is proudly worn at the news conference. (Photo: Associated Press)

Next to fly: Mercury astronaut Virgil ('Gus') Grissom. (Photo: NASA)

References

1. French, Francis and Colin Burgess, *Into That Silent Sea: Trailblazers of the Space Era, 1961-1965*, University of Nebraska Press, Lincoln, NE, 2007
2. Interview with Dee O'Hara conducted by Colin Burgess and Francis French, San Diego, CA, 18 January 2003
3. The *St. Joseph News-Press* (Missouri) newspaper article, "Shepard's 'Perfect Flight' Envy of Fellow Astronauts," 6 May 1961
4. *Ibid*
5. Interview with Dee O'Hara conducted by Colin Burgess and Francis French, San Diego, CA, 18 January 2003.
6. Glenn, John, *John Glenn: A Memoir*, Bantam Books, New York, NY, 1999, pg. 239
7. The *Telegraph-Herald* (Dubuque, Iowa) newspaper article, "Shepard Passes All Tests With Flying Colors – Doctors," Sunday 7 May 1961, pg. 3
8. Jackson, Carmault B. Jr., M.D., William K. Douglas, M.D., James F. Culver, M.D., George Ruff, M.D., Edward C. Knoblock, Ph.D. and Ashton Graybiel, M.D., "Results of Preflight and Postflight Medical Examinations," from *Results of the First U.S. Manned Suborbital Flight*, NASA, Washington, D.C., 6 June 1961
9. The *St. Joseph News-Press* (Missouri) newspaper article, "American Space Effort Belittled by Russians," 6 May 1961
10. The *Lawrence Daily Journal* (Kansas) newspaper article, "Prestige Hiked Materially by Successful Shot," 6 May 1961

11. Kraft, Chris, *Flight: My Life in Mission Control*, Dutton Books, New York, NY, 2001, pg. 142
12. The *Bonham Daily Favorite* newspaper (Bonham, Texas) article, "Shepard Gets Kiss From Wife, Medal From JFK," 8 May 1961, pg. 1
13. John F. Kennedy: "Remarks at the Presentation of NASA's Distinguished Service Medal to Astronaut Alan B. Shepard," May 8, 1961. Online by Gerhard Peters and John T. Woolley, The American Presidency Project, *http://www.presidency.ucsb.edu/ws/index.php?pid=8119*
14. The *Times-News* newspaper (Twin Falls, Idaho) article, "Astronaut Shepard is Cool Customer in Outer Space and at U.S. Capitol," 9 May 1961, pg. 6
15. The *Spartanburg Herald-Journal* newspaper (Spartanburg, South Carolina) article, "First American in Space Looks Back," 28 April 1986, pgs. D2/3
16. Shepard, Alan and Deke Slayton, with Jay Barbree and Howard Benedict, *Moon Shot: The Inside Story of America's Race to the Moon*, Virgin Books, London, U.K., 1994, pg. 131
17. Shepard, Alan, "Shepard Looks Back on Space" (written for the Associated Press), *Daily News* newspaper (Bowling Green, Kentucky), 11 May 1986
18. *The Palm Beach Post* newspaper (Florida) article, "Man of Space Wins Plaudits of Capitol," 9 May 1961, pg. 1

8

Epilogue

From the outset, the *Freedom 7* spacecraft was never intended to serve any practical purpose after its history-making flight, let alone fly into space again. Instead, it was gifted to the American people by NASA, to be preserved in a museum environment and openly exhibited for everyone to visit.

Sadly, the Soviet Union was not quite as kind to its flown manned spacecraft. In one instance the *Vostok 2* capsule, flown by Gherman Titov in making history's first day-long space flight, was converted and used as a training vessel for the upcoming Voskhod human space flight program. During a failed test for a new soft-landing parachute system, the capsule struck the ground so hard that it was crushed beyond repair. There would appear to be no indication of what happened to the remains, but there are lingering fears that it may have been unceremoniously scrapped.

Unlike his pioneering spacecraft, Alan Shepard would eventually fly into space a second time. In 1971 he commanded the Apollo 14 mission, fulfilling his long-held dream of walking on the surface of the Moon.

More than five decades on, the smaller spacecraft and the man who flew in it have both entered the history books for what they accomplished on 5 May 1961.

THE SPACECRAFT

Following its return by helicopter to Cape Canaveral, the *Freedom 7* spacecraft was subjected to several days of minute examination by engineers and technicians. It was then released for a pre-scheduled tour abroad, ahead of being placed on permanent display in the United States. On 25 May, just three weeks after it had been recovered from the Atlantic, *Freedom 7* went on display at the 24th International Aeronautical Show in Paris, France. By the end of the show on 4 June, some 650,000 fascinated attendees had taken the opportunity to view the spacecraft up close.

Crowds flocked to see *Freedom 7* on display at the Paris Air Show in 1961. (Photo: University of Central Florida)

From Paris *Freedom 7* was shipped to Italy, where it was on display from 13-25 June at the Rassegna International Electronic and Nuclear Fair in Rome. Amazingly, it drew more visitors than in Paris, with around 750,000 people lining up to inspect it. The spacecraft was then returned to the United States to undergo intensive study at the Langley Research Center in Hampton, Virginia. After that it was returned to its maker, the McDonnell Aircraft Company in St. Louis, Missouri, to be taken apart, inspected, reconstructed, and prepared for public exhibition in the National Air and Space Museum (NASM) of the Smithsonian Institution in Washington, D.C.

Four months later, on 23 October 1961, *Freedom 7* was officially presented to the Smithsonian by NASA Administrator James Webb. In his presentation speech, Webb declared, "To Americans seeking answers, proof that man can survive in the hostile realms of space is not enough. A solid and meaningful foundation for public support and the basis for our Apollo man-in-space effort is that U.S. astronauts are going into space to do useful work in the cause of all their fellow men." [1] *Freedom 7* was placed on public display in the Quonset Hut – or Air Museum Building – in the South Yard Restrictions of the National Air and Space Museum.

East met West in May 1962 when *Vostok 2* cosmonaut Gherman Titov paid an official visit to the United States. Accompanied by Mercury astronaut John Glenn, who had by then accomplished America's first manned orbital flight, and a veritable cavalcade of official vehicles and press photographers, Titov was shown some of the sights around the nation's capital, one highlight being a brief visit to the National Air and Space Museum where the cosmonaut inspected the *Freedom 7* spacecraft.

In 1965, due to keen interest abroad and through the courtesy of the Smithsonian Institution, the *Freedom 7* spacecraft was temporarily loaned to the Science Museum in Kensington, London, for a five-month exhibition. It was shipped from New York to London on the Cunard-Anchor liner *Sidonia* and delivered amid great fanfare on 17 September. The exhibition (which was advertised as lasting from 5 October 1965 to 28 February 1966) proved to be extremely popular. By the end of February it had been visited by 110,000 people. Among the visitors were Her Majesty the Queen and H.R.H. the Duke of Edinburgh, who viewed *Freedom 7* on 10 November [2].

Within the spacecraft was a lifelike model of Alan Shepard lying on his back as if preparing for liftoff, his left hand grasping the abort handle ready to fire the escape tower in the event of a mishap.

Due to great public interest, the Smithsonian agreed to an extension of the loan, allowing the exhibition to remain open until 1 May 1966. Eventually, the spacecraft was viewed by 356,000 visitors. On 18 May it was transferred to the Royal Scottish Museum in Edinburgh, where it was exhibited for several weeks in conjunction with a public talk by John Glenn on 3 June. After the exhibit was closed on 11 September, the spacecraft remained in the museum out of public view for a further three weeks to accommodate a visiting Smithsonian dignitary. Overall, the Edinburgh exhibition was seen by in excess of 200,000 visitors, this number having been collected by the "electric eye" of the museum [3].

Following its overseas sojourn, *Freedom 7* was returned to the United States and placed back on public display at the Smithsonian, where it would remain for the next 32 years.

In December 1998 the spacecraft was out on lengthy loan once again, this time to the U.S. Naval Academy in Annapolis, Maryland. The exhibition had been mounted

Alan Shepard peers into *Freedom 7* at the National Air and Space Museum of the Smithsonian Institution in Washington, D.C. (Photo: Smithsonian Institution)

America's first manned spacecraft held a great fascination for young and old alike. (Photo: Smithsonian Institution)

to honor the memory of pilot Alan Shepard, who died earlier that year. Shepard had graduated from the Academy in 1945. *Freedom 7* would remain on public display at the Armel-Leftwich Visitor Center for 14 years, honored with a place in the rotunda leading to the exhibit area. During this time, it was encased in acrylic Plexiglas and had its periscope deployed.

On 18 January 2012 the Naval Academy announced that *Freedom 7* would soon be moved from Maryland to Massachusetts and placed on temporary exhibition until December 2015 in a space gallery at the John F. Kennedy Presidential Library and Museum in Columbus Point, Boston. The spacecraft's public debut on 12 September 2012 was to coincide with the 50th anniversary of the "We choose to go to the Moon" speech that Kennedy famously delivered at Rice University in Houston in 1962.

Prior to the spacecraft taking up temporary residence in Boston, there was a major problem to resolve. This fell to experts at the Kansas Cosmosphere and Space Center in Hutchinson, Kansas, which, in consultation with the Smithsonian, developed and

Alan Shepard and John Glenn with cosmonaut Gherman Titov at the Soviet Embassy in Washington, D.C. (Photo: United Press International)

built a special cradle for exhibiting *Freedom 7* in Boston. The cradle was constructed using steel that had been washed and sandblasted in order to remove any corrosion. It was then covered with a clear protectant and painted with rubber padding where it would support the spacecraft. Jim Remar, president and chief operating officer at the Cosmosphere, said the discarded acrylic cover had been a less than ideal means of preserving the historic artifact. "The acrylic prevented the spacecraft from breathing. As materials deteriorate, they emit gas. The acrylic trapped the off-gassing in the spacecraft and [this] could accelerate or increase the rate of deterioration. With the removal of the acrylic, it is now able to breathe and the off-gas is exhausted out." [4]

However, *Freedom 7*'s journey will not end there. In 2016 the Smithsonian plans to display it as part of a major, brand new Apollo-themed gallery that tells through displays the monumental story of the Mercury, Gemini and Apollo programs.

THE ASTRONAUT

Inevitably, there are times in a nation's history when its hopes, fears and confidence in its own destiny appear to hinge on the fate of a single person. One such moment occurred on the Sun-drenched Florida spring morning of 5 May 1961, when a

Freedom 7 being delivered to the Science Museum in London, starting a year-long visit to the United Kingdom. (Photo: Science Museum/Science and Society Picture Library)

The spacecraft attracts curious onlookers outside the museum doors. (Photo: Science Museum/Science and Society Picture Library)

37-year-old test pilot squeezed into the tiny Mercury capsule named *Freedom 7*, ready to ride a rocket into the beckoning skies. Navy Cdr. Alan Shepard was trained to the hilt and fully ready to become the first American into space.

Since his selection as one of the seven Mercury astronauts in 1959, Shepard had relentlessly pursued the honor of being first. Despite this, a hollow feeling pervaded his excitement. Whatever accolades he might receive later that day, they would never make up for what he deemed to be an even greater glory. Renowned for his cocksure determination and his wicked sense of humor, he had pressed himself to the limit to be the first person to fly into space, but to his chagrin he fell just 23 days short of this prized niche in history because it went to a beaming Soviet cosmonaut named Yuri Gagarin.

Despite his Mercury flight, Alan Shepard felt somewhat relegated in history, not only as the second person to fly into space, but because his had been an all-too-brief 15-minute ballistic flight. The pioneering Mercury astronaut was demonstratively far from satisfied with the acclaim heaped on him as the first American to fly into space. He wanted something more: he wanted to fly into space again, and if determination counted for anything, then one day Alan Bartlett Shepard, Jr. would proudly stand on the Moon.

Freedom 7 on display in the museum. (Photo: Science Museum/Science and Society Picture Library)

AN AUTHENTIC AMERICAN HERO

Prophetically, Shepard called his flight aboard *Freedom 7* "just the first baby step aiming for bigger and better things," but it always galled him that an overdose of caution had cost America (and him in particular) the opportunity to be first in space [5]. His suborbital flight might seem inconsequential when compared with today's space flights, but at that time it galvanized and united Americans, giving them a renewed sense of pride and accomplishment. It also set in motion mankind's most audacious scientific undertaking. Just twenty days after Shepard's triumphant return to Earth, President Kennedy stood before Congress and challenged his nation to land a man on the Moon before the decade was out.

After fellow Mercury astronaut Gus Grissom had virtually replicated Shepard's flight with a second ballistic flight in July, NASA decided to press on with orbital missions. This was first achieved by John Glenn on board *Friendship 7* in February 1962. After two further manned orbital flights by Scott Carpenter and Wally Schirra, it was announced that Gordon Cooper would wrap up the Mercury project with a 22-orbit flight in May 1963.

Freedom 7 is shown here after its safe arrival at the Royal Scottish Museum. (Photo: The Scotsman Publications Ltd.)

In 1998, following the death of former graduate Alan Shepard, the spacecraft went on long-term display at the U.S. Naval Academy, Annapolis, Maryland. (Photo: U.S. Naval Academy)

Following its arrival in Boston, Massachusetts in 2012, *Freedom 7* was delivered to the JFK Presidential Library and Museum as a temporary exhibition. (Photo credit: Rick Friedman, JFK Library Foundation)

A smiling Alan Shepard in training for his MR-3 mission. (Photo: NASA)

However, Alan Shepard was keen to fly again, and if it meant using a little of his renowned tenacity then he was prepared to give it his best shot. He knew a spacecraft designated 15B had already been manifested to a possible final Mercury mission and it had been substantially upgraded, making it capable of operating a prolonged flight. Since he was Cooper's backup and his colleagues were now engaged in assignments specifically related to the Gemini and Apollo projects, he would automatically be the prime pilot for an additional flight, if one were to occur. Shepard strenuously argued for such a mission, even renaming spacecraft 15B *Freedom 7 II*, and having that logo painted on its exterior. As NASA was lukewarm to the idea, in a typically audacious move Shepard went around his bosses in the space agency and attempted to enlist the personal support of President Kennedy, who told him that the decision would rest with NASA Administrator James Webb.

Webb carefully weighed up all the options, and when he stood before the Senate Space Committee in June 1963 he began by stating, in part, "There will be no more Mercury shots." He went on to explain that Project Mercury had now satisfactorily accomplished its goals, and there should be new priorities. All the energies of NASA and its contractors, he said, should now be fully employed in focusing on the Gemini and Apollo missions. As it turned out, even if Shepard had realized his goal of being assigned a second one-man flight, it was a mission he would never have been able to fly.

An early consolation came when Shepard was selected to fly the first Gemini two-man mission, with rookie astronaut Tom Stafford as his copilot. Shortly after starting preliminary training in the simulators in early 1964, Shepard was suddenly struck by an ailment which threatened to end not only his astronaut career, but also his days as a pilot. He awakened one morning feeling slightly giddy, and upon trying to stand up he collapsed. Thinking it to be an isolated incident, he was not overly concerned. But five days later he suffered a second sudden bout of dizziness, and this time began to

Capsule 15B, unofficially named *Freedom 7 II*, shown in its orbital configuration at the Udvar-Hazy Center of the National Air and Space Museum in Washington, D.C. (Photo courtesy of Stéphane Sebile)

The unofficial logo *Freedom 7 II* painted on the side of Spacecraft 15B at the request of Alan Shepard. (Photo courtesy of Stéphane Sebile)

vomit uncontrollably. This incident left him with a loud, recurring ringing in his left ear. After these attacks had struck him down several times, Shepard finally realized it was not something he could simply tough out, and made an appointment with the flight surgeons. After extensive tests, a panel of NASA doctors recommended he be removed immediately from his flight assignment.

The ailment proved to be Ménière's Syndrome. "The problem is not considered very significant for an Earth-bound person, but it sure can finish you as a pilot," he said during a 1970 interview for *Naval Aviator News*. "I convinced myself it would eventually work itself out, but it didn't. Tom Stafford had told me about Dr. House, out in Los Angeles, who could perform an operation on this particular kind of inner ear trouble. At first it sounded a little risky, but in 1968 I finally decided on having it done. With NASA's permission I went out to California. In order to keep the whole business quiet, Dr. House and I agreed that I should check into the hospital under an assumed name. It was the doctor's secretary who came up with it. So, as Victor Poulis, I had the operation, and six months later my ear was fine." [6]

FLY ME TO THE MOON

Although his surgery was successful, Shepard had lost his chance to fly on Gemini, and there were serious doubts that he would ever fly into space again. To remain part of the astronaut cadre, he had earlier accepted an interim appointment as Chief of the Astronaut Office, giving him a major influence in the training and assignment of his fellows. Eventually, his never-say-die attitude would see him regain active astronaut status, and he promptly launched a determined campaign for a slot aboard a manned Apollo lunar mission.

The Apollo 14 crew of Stuart Roosa, Alan Shepard, and Edgar Mitchell. (Photo: NASA)

Almost a decade after his historic flight aboard *Freedom 7*, Shepard was launched into space for a second and final time on 31 January 1971 as the commander of the Apollo 14 mission. Aged 47, he became the oldest of the twelve men to place their boot prints in the lunar dust. Along with Edgar Mitchell, he spent 33 hours exploring the Fra Mauro terrain. He freely admits that when he stepped off the Lunar Module *Antares* for the first time and stood on the lunar surface he shed tears of wonderment and joy.

At the end of their final excursion, Shepard impishly pulled out a club head which he had secretly brought along, and clipped it onto the long handle of a tool. He then dropped a golf ball onto the surface and attempted a modified one-armed backswing. "Unfortunately, the suit is so stiff I can't do this with two hands," he reported back to Earth, "but I'm going to try a little sand trap shot here." Using only his right hand he whacked the first of two balls for a distance he later said with a broad aviator's grin was "miles and miles."

Alan Shepard stands on the surface of the Moon. (Photo: NASA)

A SPACE FLIGHT LEGEND REMEMBERED

In the wake of his Apollo mission Alan Shepard was promoted to the rank of rear admiral, becoming the first astronaut to achieve such status. He resigned from both NASA and the Navy in 1974. After his Mercury flight in 1961 he had been awarded the Distinguished Flying Cross and NASA's Distinguished Service Medal, and with his resignation from the Navy he also added with pride the Congressional Space Medal of Honor.

Post-NASA, Shepard followed in his father's footsteps by venturing into banking, real estate and investments, and other private business, in the process making himself a considerable fortune. He also dabbled on the fringe of politics by joining the board of the right-wing Freedom Forum in 1993.

In 1984, he joined with the other five surviving Mercury astronauts in setting up the Mercury 7 Foundation, a science and engineering scholarship fund for college students, and served as its founding president. Today, under the revised name of the Astronaut Scholarship Foundation, it pursues the same goals.

Once, in an interview for the Hall of Science and Exploration, Shepard was asked for his proudest accomplishment, which he said was being chosen to make the first manned American flight into space. "That was competition at its best," he explained, with his usual unapologetic candor. "Not because of the fame or the recognition that went with it, but because of the fact that America's best test pilots went through this selection process down to seven guys, and of those seven, I was the first one to go. That will always be the most satisfying thing for me.

"During the actual process of flying aircraft, or flying the *Spirit of St. Louis*, one doesn't think of oneself as being a hero or historical figure. One does it because the

At the Pentagon on 26 August 1971, a proud Alan Shepard is awarded the shoulder boards of a rear admiral's rank by Navy Secretary John L.H. Chafee (left) and Adm. Ralph W. Cousins, Vice Chief of Naval Operations. (Photo: Associated Press)

challenge is there, and one feels reasonably qualified to accomplish it." After a pause he added, "I must admit, maybe I am a piece of history after all." [7]

On Tuesday, 21 July 1998, the world lost America's first astronaut in space to the insidious disease leukemia. He had fought a typically stoic and mostly private two-year battle against this cancer, but it was a fight even he could not win. R/Adm Alan Shepard, an authentic twentieth-century hero, passed away peacefully in his sleep at the Monterey Community Hospital in California. He was 74 years old.

Biographer Neal Thompson says Shepard's whole life was about competition. "Whether it was in sports as a youth, or competing among other naval aviators when he was a carrier pilot, and then it just sort of ramped up at each stage of his career, becoming a test pilot where he competed with some of the best aviators on the planet and then to be selected among this extremely elite group of Mercury 7 astronauts and then to compete against them for that first ride. But I think he thrived on that and it was fun to explore what that meant in the scope of the space program." [8]

On 25 August, barely a month after the loss of her husband, Louise Shepard died of a heart attack while on a flight from San Francisco to her home in Monterey. She was returning from Colorado after visiting one of her daughters.

Alan and Louise Shepard were cremated and their ashes committed to the sea in Stillwater Cove near Pebble Beach, California. A small memorial stone for both was placed in the Forest Hill Cemetery in Derry, New Hampshire. They are survived by daughters Alice Wackermann, Julie Jenkins, and Laura Churchley, plus their six grandchildren.

Alan B. Shepard, Jr., the first American in space and Apollo 14 moonwalker. (Photo: NASA)

The memorial stone for Alan and Louise Shepard. (Photo courtesy of David Lee Tiller)

References

1. Presentation speech by NASA Administrator James Webb, 23 October 1961. From Encyclopedia Astronautica website at *http://www.astronautix.com/Astros/webb.htm*
2. Letter from D.H. Follett, Director of the Science Museum, London, to S. P. Johnston, Director, National Air and Space Museum, Smithsonian Institution, Washington, D.C. Letter dated 2 June 1966
3. Email correspondence with Dr. Tacye Phillipson, Senior Curator, National Museums Scotland, 18-21 December 2012
4. Stephens, Ken, article, "Space Works done with cradle for Freedom 7," *The Hutchinson News* (Kansas), 10 Sep. 2012
5. Dunn, Marcie, *Los Angeles Times* newspaper article, "Alan Shepard Recalls Pioneering Flight," issue 28 April 1991
6. *Naval Aviation News* article, "Once a Fighter Pilot," by Cdr. Ted Wilbur, issue November 1970
7. *Academy of Achievement* article, "Admiral Alan Shepard, Jr.: Pioneer of the Space Age." Uncredited interview conducted on 1 February 1991, Houston, Texas. Website: http://www.achievement.org/autodoc/page/she0int-1
8. Atkinson, Nancy, *Alan Shepard: Complicated, Conflicted and the Consummate Astronaut*, interview article with Neal Thompson for *Universe Today*, 5 May 2011. Website: http://www.universetoday.com

Appendix 1: In his own words

Pilot's flight report by Alan B. Shepard, Jr.

Taken from the NASA paper (in conjunction with the National Institutes of Health and the National Academy of Sciences): *Proceedings of a Conference on Results of the First U.S. Manned Suborbital Flight*, 6 June 1961, Washington, D.C.

(Most references by Shepard to images screened during his presentation deleted)

INTRODUCTION

My intention is to present my flight report in narrative form and to include three phases. These phases shall be: (1) the period prior to launch, (2) the flight itself, and (3) the post-flight debriefing period. I intend to describe my feelings and reactions and to make comments pertinent to these three areas. I also have an onboard film of the flight to show at the end of my presentation.

PRE-FLIGHT PERIOD

Astronaut D.K. Slayton in a previous paper described the program followed by the Project Mercury astronauts during a two-year training period with descriptions of the various devices used. All of these devices provided one thing in common: namely, the feeling of confidence that the astronauts achieved from their use. Some devices, of course, produced more confidence than others but all were very well received by the group. There are three machines or training devices which provided the most assistance. The first of these is the human centrifuge. We used the facilities of the U.S. Naval Air Development Center in Johnsville, Pennsylvania, which provided the centrifuge itself and a computer to control its inputs. This computer, through an instrument display, provided a control task similar to that of the Mercury spacecraft, with inputs of the proper aerodynamic and moment-of-inertia equations. Thus, we were able to experience the acceleration environment while simultaneously controlling the spacecraft on a simulated manual system. This experience gave us the feeling of muscle control for circulation and breathing, transmitting, and general control of the spacecraft. I found that the flight environment was very close to the environment provided by the centrifuge. The flight accelerations were smooth, of the same magnitude used during training, and certainly in no way disturbing.

The second training device that proved of great value was the procedures trainer. This device will be recognized as an advanced type of the Link trainer, which was used for instrument training during the last war. We were able to use it to correlate pre-flight planning, to practice simulated control maneuvers, and to practice operational techniques. The Space Task Group has two such trainers, one at Langley Field, Virginia, the other at Cape Canaveral, Florida, and both are capable of the simultaneous training of pilots and ground crews. As a result of the cross-training between pilots and the ground crews at the Project Mercury Control Center, we experienced no major difficulties during the flight. We had learned each other's problems and terminology, and I feel that we have a valuable training system in use for present and for future flights.

The third area of pre-flight training, which is considered as one of importance, concerns working with the spacecraft itself. The Mercury spacecraft is tested at Cape Canaveral before being attached to the Redstone launch vehicle. These tests provide an excellent opportunity for pilots to learn the idiosyncrasies of the various systems. After the spacecraft has been placed on the launch vehicle, more tests are made just prior to launch day. The pilots have a chance to participate in these tests and to work out operational procedures with the blockhouse crew.

These three areas then, the centrifuge, the procedures trainer, and spacecraft testing at the launching area, provided the most valuable aids during the training period. We spent two years in training, doing many things, following many avenues in our desire to be sure that we had not overlooked anything of importance. As a general comment concerning future training programs, these experiences will undoubtedly permit us to shorten this training period.

During the days immediately preceding the launch, the pre-flight physicals were given. These examinations do not involve more than the usual profiling, listening, and other medical tests, but I hope that fewer body fluid examples are required in the future. I felt as though an unusual number of medics were used.

Pre-flight briefing was held at 11 a.m. on the day before launch to correlate all operational elements. This briefing was helpful since it gave us a chance to look at weather, radar, camera, and recovery force status. We also had the opportunity to review the control procedures to be used during flight emergencies as well as any late inputs of an operational nature. This briefing was extremely valuable to me in correlating all of the details at the last minute.

PERIOD OF FLIGHT

I include as part of the flight period the time from insertion into the spacecraft on the launching pad until the time of recovery by the helicopter. The voice and operational procedures developed during the weeks preceding the launch were essentially sound.

The countdown went smoothly, and no major difficulties were encountered with the ground crews, the control-central crew, and the pilot. There has been some comment in the press about the length of time spent in the spacecraft prior to launch, some 4 hours 15 minutes to be exact. This period was about two hours longer than had been planned. A fact that is most encouraging is that during this time there was no significant change in pilot alertness and ability. The reassurance gained from this experience applies directly to our upcoming orbital flights, and we now approach them with greater confidence in the ability of the pilots, as well as in the environmental control systems.

Our plan was for the pilot to report to the blockhouse crew primarily prior to the T-2 minutes on hard wire circuits, and to shift control to the Center by use of radio frequencies at T-2 minutes. This shift worked smoothly and continuity of information to the pilot was good. At lift-off I started a clock timer in the spacecraft and prepared for noise and vibration. I felt none of any serious consequence. The cockpit section experienced no vibration and I did not even have to turn up my radio receiver to full volume to hear the radio transmissions. Radio communication was verified after lift-off, and then periodic transmissions were made at 30-second intervals for the purpose of maintaining voice contact and of reporting vital information to the ground.

Some roughness was expected during the period of transonic flight and of maximum dynamic pressure. These events occurred very close together on the flight, and there was general vibration associated with them. At one point some head vibration was observed. The degradation of vision associated with this vibration was not serious. There was a slight fuzzy appearance of the instrument needles. At T+1 minute 21 seconds I was able to observe and report the cabin pressure without difficulty. I accurately described the cabin pressure as "holding at 5.5 p.s.i.a." The indications of the various needles on their respective meters could be determined accurately at all times. We intend to alleviate the head vibration by providing more foam rubber for the head support and a more streamlined fairing for the spacecraft adapter ring. These modifications should take care of this problem for future flights.

I had no other difficulty during powered flight. The training in acceleration on the centrifuge was valid, and I encountered no problem in respiration, observation, and reporting to the ground.

Rocket cutoff occurred at T+2 minutes 22 seconds at an acceleration of about 6 g. It was not abrupt enough to give me any problem and I was not aware of any uncomfortable sensation. I had one switch movement at this point which I made on schedule. Ten seconds later, the spacecraft separated from the launch vehicle, and I was aware of the noise of the separation rockets firing. In another 5 seconds the periscope had extended and the autopilot was controlling the turnaround to orbit attitude. Even though this test was only a ballistic flight, most of the spacecraft action and piloting techniques were executed with orbital flight in mind. I would like to make the point again that attitude control in space differs from that in conventional aircraft. There is a penalty for excessive use of the peroxide fuel and we do not attempt to control continually all small rate motions. There is no aerodynamic damping in space to prevent attitude deviation, but neither is there any flight-path excursion or acceleration purely as a function of variation in spacecraft angles.

At this point in the flight I was scheduled to take control of the attitude (angular position) by use of the manual system. I made this manipulation one axis at a time, switching to pitch, yaw and roll in that order until I had full control of the craft. I used the instruments first and then the periscope as reference controls. The reaction of the spacecraft was very much like that obtained in the air-bearing trainer (ALFA trainer) described previously in the paper by Astronaut Slayton. The spacecraft movement was smooth and could be controlled precisely. Just prior to retrofiring, I used the periscope for general observation.

The particular camera orientation during my flight happened to include many clouds and is not as clear for land viewing. This photograph shows the contrast between land and water masses, the cloud cover and its effect, and a good view of the horizon. There appears to be a haze layer at the horizon. This haze is a function not only of particles of dust, moisture, and so forth, but also of light refraction through atmospheric layers. The sky itself is a very deep blue, almost black, because of the absolute lack of light-reflecting particles. We are encouraged that the periscope provides a good viewing device as well as a backup attitude-control indicator and navigation aid.

At about this point, as I have indicated publicly before, I realized that somebody would ask me about weightlessness. I use this example again because it is typical of the lack of anything upsetting during a weightless or zero-g environment. Movements, speech, and breathing are unimpaired and the entire sensation is most analogous to floating. The NASA intends, of course, to investigate this phenomenon during longer periods of time, but the astronauts approach these periods with no trepidation.

Control of attitude during retrofiring was maintained on the manual system and was within the limits expected. There was smooth transition from zero gravity to the thrust of the retrorocket and back to weightless flying again. After the retrorockets had been fired, the automatic sequence acted to jettison them. I could hear the noise and could see one of the straps falling away in view of the periscope. My signal light inside did not show proper indication so I used the manual backup control and the function indicated proper operation.

After retrorockets were jettisoned, I used a combination of manual and electric control to put the spacecraft in the reentry attitude. I then went back to autopilot control to allow myself freedom for some other actions. The autopilot control functioned properly so I made checks on the high frequency voice link for propagation characteristics and then returned to the primary UHF voice link. I also looked out both portholes to get a general look at the stars or planets as well to get oblique horizon views. Because of the sun angle and light levels I was unable to see any celestial bodies. The Mercury Project plans are to investigate these phenomena further on later flights.

At an altitude of about 200,000 feet, or at the edge of the sensible atmosphere, a relay was actuated at 0.05 g. I had intended to be on manual control for this portion of the flight but found myself a few seconds behind. I was able to switch to the manual system and make some controlling motions during this time. We feel that programming for this maneuver is not a serious problem and can be corrected by allowing a little more time prior to the maneuver to get ready. We were anxious to get our money's worth out of the flight and consequently we had a full flight plan. However, it paid off in most cases as evidenced by the volume of data collected on pilot actions.

The reentry and its attendant acceleration pulse of 11 g was not unduly difficult. The functions of observation, motion, and reporting were maintained, and no respiration difficulties were encountered. Here again, the centrifuge training had provided good reference. I noticed no loss of peripheral vision, which is the first indication of "gray out."

After the acceleration pulse I switched back to the autopilot. I got ready to observe parachute opening. At 21,000 feet the drogue parachute came out on schedule as did the periscope. I could see the drogue and its action through the periscope. There was no abrupt motion at drogue deployment. At 10,000 feet the main parachute came out and I was able to observe the entire operation through the periscope. I could see the streaming action as well as the unreefing action and could immediately assess the condition of the canopy. It looked good and the opening shock was smooth and welcome. I reported all of these events to the control center and then proceeded to get ready for landing.

I opened the faceplate of the helmet and disconnected the hose which supplies oxygen to its seal. I removed the chest strap and the knee restraint straps. I had the lap belt and shoulder harness still fastened. The landing did not seem any more severe than a catapult shot from an aircraft carrier. The spacecraft hit and then flopped on its side so that I was on my right side. I felt that I could immediately execute an underwater escape should it become necessary. Here again, our training period was giving us dividends. I could see the water covering one porthole. I could see the yellow dye marker out the other porthole and, later on, I could see one of the helicopters through the periscope.

The spacecraft righted itself slowly and I began to read the cockpit instruments for data purposes after impact. I found very little time for that since the helicopter was already calling me. I made an egress as shown in the training movie; that is, I sat on the edge of the door sill until the helicopter sling came my way .... The hoist itself was uneventful. At this point, I would like to mention a device that we use on our pressure suits that gives watertight integrity. There is a soft rubber cone attached to the neck ring seal of the suit. When the suit helmet is on, this rubber is rolled and stowed below the lip of the neck ring seal bearing. With the helmet off, this collar or neck cone is rolled up over the bearing and against the neck of the pilot where it forms a watertight seal. The inlet valve fitting has a locking flapper valve. Thus the suit is waterproof and provides its own buoyancy.

POST-FLIGHT DEBRIEFING

The helicopter took me to the aircraft carrier *Lake Champlain*, where the preliminary medical and technical debriefing commenced. Since no serious physiological defects were noted, only an immediate cursory examination was necessary. The period I spent in talking into a tape recorder at this time with the events fresh in my mind was also a help. I had a chance to report before becoming confused with the "facts."

I went from the carrier to the Grand Bahama Island where I spent the better part of two days in combined medical and technical debriefings. A great deal of data was gathered, and the experience was not unduly uncomfortable. It appears profitable to provide a location where a debriefing of this sort can be accomplished.

It is now our plan to show you a film of the flight taken from the onboard equipment. The film has been taken from the onboard camera and step-printed to real time, and the tape recorder conversations have been synchronized for the entire flight. (*At this point onboard footage was shown*)

Appendix 1: In his own words

In closing I would like to say that the participants in Project Mercury are indeed encouraged by the pilot's ability to function during the ballistic flight which has just been described. No inordinate physiological change has been observed, and the control exercised before and after the flight overwhelmingly support this conclusion. The Space Task Group is also encouraged by the operation of the spacecraft systems in the automatic mode, as well as in the manual mode. We are looking forward to more flights in the future, both of the ballistic as well as the orbital type.

Appendix 2: NASA-released transcript of voice communications during MR-3 flight between spacecraft *Freedom 7* (Alan Shepard) and CapCom (Deke Slayton) in the Mercury Control Center

Launch communication beginning at minus 60 seconds:	
−00.01.00 (CapCom):	One minute and counting. Mark.
−00.00.50 (Shepard):	*Roger.*
−00.00.45 (CapCom):	Forty-five and counting. Mark.
−00.00.40 (Shepard):	*Roger.*
−00.00.30 (CapCom):	Firing command, 30. Mark.
−00.00.25 (Shepard):	*Roger ... Periscope has retracted.*
−00.00.28 (CapCom):	That is the best periscope we've got.
−00.00.20 (Shepard):	*Main bus 24 volts, 26 amps.*
−00.00.15 (CapCom):	15 ... 10, 9, 8, 7, 6, 5, 4, 3, 2, 1, zero. Liftoff.

After liftoff:	
+00.00.02 (Shepard):	*Ah, Roger. Liftoff, and the clock has started.*
+00.00.05 (CapCom):	Okay, José, you're on your way.
+00.00.08 (Shepard):	*Roger. Reading you loud and clear.*
+00.00.13 (CapCom):	So can I you.
+00.00.25 (Shepard):	*This is Freedom Seven. The fuel is go, 1.2 g, cabin at 14 psi, oxygen is go.*
+00.00.32 (CapCom):	Understand.
+00.00.48 (Shepard):	*Freedom Seven is still go.*
+00.00.58 (Shepard):	*This is Seven. Fuel is go, 1.8 psi cabin, and the oxygen is go.*
+00.01.21 (Shepard):	*Cabin Pressure is holding at 5.5. Cabin holding at 5.5.*
+00.01.27 (CapCom):	I can understand. Cabin holding at 5.5.
+00.01.33 (Shepard):	*Fuel is go, 2.5 g, cabin 5.5, oxygen is go, the main bus is 24, and the isolated battery is 29.*
+00.01.42 (CapCom):	Rog. Reading 5.5. Trajectory looks good.
+00.01.50 (Shepard):	*Okay, it's a lot smoother now. A lot smoother.*
+00.01.56 (CapCom):	Very good.

(continued)

Appendix 2: NASA-released transcript of voice communications...

(continued)

After liftoff:

+00.02.01 (Shepard):	*Seven here. Fuel is go, 4 g, 5.5 cabin, oxygen go. All systems are go.*
+00.02.09 (CapCom):	All systems go. Trajectory okay.
+00.02.15 (Shepard):	*5 g.*
+00.02.22 (Shepard):	*Cutoff. Tower jettison green.*
+00.02.05 (CapCom):	Roger.
+00.02.27 (Shepard):	*Disarm.*
+00.02.32 (Shepard):	*Cap sep is green.*
+00.02.34 (CapCom):	Cap sep comes up.
+00.02.35 (Shepard):	*Periscope is coming out and the turnaround has started.*
+00.02.41 (CapCom):	Roger.
+00.02.50 (Shepard):	*ASCS is okay.*
+00.02.53 (Shepard):	*Control movements.*
+00.02.54 (CapCom):	Roger.
+00.03.04 (Shepard):	*Okay, switching to manual pitch.*
+00.03.08 (CapCom):	Manual pitch.
+00.03.21 (Shepard):	*Pitch is okay.*
+00.03.24 (Shepard):	*Switching to manual yaw.*
+00.03.29 (CapCom):	I can understand. Manual yaw.
+00.03.35 (CapCom):	Okay.
+00.03.42 (Shepard):	*Yaw is okay. Switching to manual roll.*
+00.03.48 (CapCom):	Manual roll.
+00.03.55 (Shepard):	*Roll is okay.*
+00.03.57 (CapCom):	Roll okay. Looks good here.
+00.03.59 (Shepard):	*On the periscope. What a beautiful view.*
+00.04.03 (CapCom):	I'll bet it is.
+00.04.05 (Shepard):	*Cloud cover over Florida. Three-to-four-tenths near the Eastern coast. Obscured up to Hatteras.*
+00.04.20 (Shepard):	*I can see Okeechobee. Identify Andros Island. Identify the reefs.*
+00.04.28 (CapCom):	Roger. Down to retro: 5, 4, 3, 2, 1, retro-angle.
+00.04.44 (Shepard):	*Start retro sequence. Retro attitude on green.*
+00.04.49 (CapCom):	Roger.
+00.04.56 (Shepard):	*Control is smooth.*
+00.05.02 (CapCom):	Roger. Understand all going smooth.
+00.05.13 (Shepard):	*There's ... Retro one. Very smooth.*
+00.05.15 (CapCom):	Roger. Roger.
+00.05.16 (Shepard):	*Retro Two.*
+00.05.23 (Shepard):	*Retro three.*
+00.05.31 (Shepard):	*All three retros are fired.*
+00.05.33 (CapCom):	All right on the button.
+00.05.35 (Shepard):	*Okay. Three retros have fired. Retro-jettison is back to armed.*
+00.05.40 (CapCom):	Roger. Do you see the booster?
+00.05.45 (Shepard):	*No. Negative.*
+00.05.55 (Shepard):	*Switching to fly-by-wire.*
+00.06.01 (CapCom):	Fly-by-wire. Understand.
+00.06.11 (Shepard):	*Roll is okay.*
+00.06.14 (CapCom):	Roger.
+00.06.16 (Shepard):	*Roger. Do not have a light.*

(continued)

Appendix 2: NASA-released transcript of voice communications... 253

(continued)

After liftoff:	
+00.06.21 (CapCom):	Understand you do not have a light.
+00.06.25 (Shepard):	*I do not have a light. I see the straps falling away. I heard a noise. I will use override.*
+00.06.29 (CapCom):	Roger.
+00.06.30 (Shepard):	*Override used. The light is green.*
+00.06.34 (CapCom):	… retroject.
+00.06.36 (Shepard):	*Ahhh, Roger. Periscope is retracting.*
+00.06.41 (CapCom):	Periscope retracting.
+00.06.49 (Shepard):	*I'm on fly-by-wire. Going to reentry attitude.*
+00.06.56 (CapCom):	Reentry attitude, Roger. Trajectory is right on the button.
+00.07.04 (Shepard):	*Okay, Buster. Reentry attitude. Switching to ASCS normal.*
+00.07.09 (CapCom):	Roger.
+00.07.14 (Shepard):	*ASCS is okay.*
+00.07.18 (CapCom):	Understand.
+00.07.25 (Shepard):	*Switching HF for radio check.*
+00.07.32 (CapCom):	Freedom Seven, CapCom. How do you read HF?
+00.07.39 (Shepard):	*Ahhh, Roger. Reading you loud and clear HF, Deke. How me?*
+00.07.44 (CapCom):	Back to UHF.
+00.08.04 (Shepard):	*This is Freedom Seven.*
+00.08.10 (Shepard):	*G buildup, 3, 6, 9.*
+00.08.21 (Shepard):	*Okay, okay.*
+00.08.23 (CapCom):	Coming through loud and clear.
+00.08.27 (Shepard):	*Okay.*
+00.08.36 (Shepard):	*Okay.*
+00.08.40 (CapCom):	CapCom; your impact will be right on the button.
+00.08.47 (Shepard):	*This is Seven. Okay.*
+00.08.51 (Shepard):	*45,000 feet now.*
+00.08.56 (Shepard):	*Aah, 40,000 feet.*
+00.08.58 (Shepard):	*I'm back on ASCS.*
+00.09.05 (Shepard):	*35,000.*
+00.09.14 (Shepard):	*30,000 feet.*
+00.09.15 (CapCom):	CapCom; how do you read now?
+00.09.18 (Shepard):	*Loud and clear. 25,000.*
+00.09.20 (Shepard):	*Aah, Roger, Deke, read you loud and clear. How me?*
+00.09.25 (CapCom):	Switching over to GBI.
+00.09.35 (Shepard):	*Aah, Roger.*

CapCom at GBI (Grand Bahama Island) takes over communications:	
+00.09.39 (Shepard):	*The drogue is green at 21[,000]. The periscope is out. The drogue is out.*
+00.09.48 (Shepard):	*Okay at drogue deploy. I've got seven zero percent auto – nine zero percent manual. Oxygen is still okay.*
+00.09.55 (GBI):	Can you read?
+00.09.57 (Shepard):	*Thirty five. Sixty seconds.*
+00.10.00 (GBI):	Can you read?
+00.10.02 (Shepard):	*I read. And the snorkel's [out] at about 15,000 feet.*

(continued)

Appendix 2: NASA-released transcript of voice communications...

(continued)

CapCom at GBI (Grand Bahama Island) takes over communications:	
+00.10.06 (Shepard):	*Emergency flow rate is on.*
+00.10.08 (Shepard):	*Standing by for main.*
+00.10.15 (Shepard):	*Main on green.*
+00.10.18 (Shepard):	*Main chute is reefed.*
+00.10.22 (Shepard):	*Main chute is green. Main chute is coming unreefed and it looks good.*
+00.10.28 (Shepard):	*Main chute is good. Rate of descent is reading about 35 feet per second.*
+00.10.40 (Shepard):	*Hello CapCom. Freedom Seven. How do you read?*
+00.10.55 (Shepard):	*Hello Cardfile 23 [recovery aircraft], this is Freedom Seven. How do you read?*
+00.11.00 (GBI):	Freedom Seven, this is Indian CapCom. Do you read me?
+00.11.03 (Shepard):	*Affirmative, Indian CapCom, let me give you a report. I'm at 7,000 feet, the main chute is good, the landing bag is on green, my peroxide has dumped, my condition is good.*
+00.11.22 (GBI):	Roger, Freedom Seven. I understand you're at 7,000 feet. Your main chute is open. Your ... is okay.
+00.11.29 (Shepard):	*That is affirmative. Please relay.*
+00.12.36 (Shepard):	*Hello Cardfile 23, Cardfile 23, Freedom Seven. Over.*
+00.12.42 (Cardfile 23):	Aah, Freedom Seven, Freedom Seven. This is Cardfile 23. Over.
+00.12.49 (Shepard):	*Aah, this is Seven. Relay back to CapCom please. My altitude now 4,000 feet, condition as before. The main chute is good, the landing bag has deployed, the periscope has dumped.*
+00.13.14 (Cardfile 23):	Aah, Rog. Understand ... relay.
+00.13.50 (Cardfile 23):	CapCom, this Cardfile 23.
+00.14.03 (Shepard):	*Aah, Cardfile 23. Freedom Seven.*
+00.14.06 (Cardfile 23):	Cardfile, this is 23.
+00.14.09 (Shepard):	*I'm about 1,000 feet now. The main chute still looks good. The rate of descent is indicating 30 feet per second.*
+00.14.15 (Slayton):	Ahh, rog.
+00.14.43 (Slayton):	Freedom Seven, this is ... transmission from ... ah ... Cape CapCom.
+00.14.59 (Shepard):	*This is Seven. Go ahead.*
+00.15.02 (Slayton):	... transmitted this time.
+00.15.05 (Shepard):	*Negative. Just relaying my condition is still good. I'm getting ready for impact.*
+00.15.22 [Splashdown]:	

Appendix 3: *Freedom 7* spacecraft pre-launch activities

Mercury Spacecraft #7, which became known as *Freedom 7*, was delivered from the McDonnell Aircraft Corporation plant in St. Louis, Missouri to Hangar S at Cape Canaveral on 9 December 1960. Upon delivery, the instrumentation system and selected items of the communication system were removed from the capsule to be bench treated. During this bench-test period, the capsule underwent rework which included the cleaning up of discrepancy items deferred from St. Louis and making changes to the capsule that were required to be made prior to beginning systems tests.

Systems test were begun as soon as all instrumentation and communications components were reinstalled in the capsule. These tests required a total of 46 days. During this period the electrical, sequential, instrumentation, communication, environmental, reaction control, and stabilization and control systems were individually tested. Included in the test of the environmental system were two runs in an altitude chamber with an astronaut installed in the capsule.

At the completion of systems tests, another work period was scheduled in which the landing bag system was installed on the capsule. Following this work period, a simulated flight test was performed, followed by the installation of pyrotechnics and parachutes. The capsule was then weighed, balanced, and delivered to the launching pad to be mated with the Redstone booster. Nineteen days were spent on the launching pad, prior to launch, testing the booster and capsule systems, both separately, and as a unit. Also, practice insertions of an astronaut into the capsule were performed during this period.

Simulated flight 1 with the booster was accomplished at the completion of systems tests on the launching pad. A change was then required in the booster circuitry which necessitated another simulated flight test (simulated flight 2). The capsule-booster combination was then ready for flight. The flight was postponed several days due to weather; however, this allowed time for replacing instrumentation components which were malfunctioning. A final simulated flight was then run (simulated flight 3). The capsule was launched two days after this final test.

Appendix 3: *Freedom 7* spacecraft pre-launch activities

MODIFICATIONS MADE

During capsule systems tests and work periods, both in Hangar S and on the launching pad, modifications were made to the capsule as a result of either a capsule malfunction or an additional requirement placed on the capsule. The most significant modifications made to Spacecraft 7 while at Cape Canaveral were as follows:

(a) Manual sensitivity control and a power cut-off switch were added to the VOX (Voice-Operated Transmitter) relay.
(b) A check valve was installed between the vacuum relief valve and the snorkel inflow valve.
(c) The cabin pressure relief valve was replaced with one which would not open until it experienced an equivalent head of 15 inches of water.
(d) Screens were added at the heat barriers upstream of the thrust chambers (downstream of the solenoid valves).
(e) The high-thrust pitch and yaw thrusters were welded at the juncture between the thrust chambers and the heat barriers.
(f) The cables to the horizon scanners in the antenna canister were wrapped with reflective tape to minimize radio-frequency (RF) interference from capsule communications components.
(g) The retro-interlock circuit was bypassed by installing a jumper plug in the amplifier-calibrator.
(h) Permission relays were installed to both the capsule-adapter ring limit switches and the capsule-tower ring limit switches.
(i) Capacitors were installed in the circuits to the orbit attitude, retro-jettison, and impact inertia arm time delay relays.
(j) Capsule wiring was changed to extend the periscope at 21,000 feet.
(k) The potting on the capsule adapter umbilical connectors was extended 0.75 inches from both connector ends and the connector was wrapped with asbestos and heat reflective tape. Also, the fairings over these connectors were cut away and a cover was added which provided more clearance between the fairings and the connectors.
(l) The lower pressure bulkhead was protected from puncture damage that might result from heat sink recontact. Aluminum honeycomb was added, bolts reversed, and brackets with sharp protrusions were potted solidly with RTV-90 and plates between the brackets and the bulkhead.
(m) Pitch indicator markings were changed from –43 to –34 degrees for retro-attitude indication

Appendix 4: Freedom 7 events and trajectory

The sequence of events on the MR-3 mission occurred according to plan, while the actual trajectory flown was very close to the nominal calculated trajectory.

In the first table below, the sequence of major events show the planned and actual times at which they occurred. The second table lists actual and planned trajectory parameters.

Event	Planned Time	Actual Time
Booster Cut-Off	02:23.1	02:21.8
Tower Release	02:23.1	02:22.0
Tower Escape Rocket Fire	02:23.1	02:22.2
Capsule Separation	02:33.1	02:32.3
Time of Retro-fire Sequence	04:41.5	04:44.7
Retro-attitude Comm. Relay	04:41.5	04:44.7
Retro #1 Fire	05:11.5	05:14.1
Retro #2 Fire	05:16.5	05:18.8
Retro #3 Fire	05:21.5	05:23.6
Retro-package Jettison	06:10.5	06:13.6
.05 g Relay	07:43.0	07:48.2
Drogue Chute Deploy	09:36.0	09:38.1
Main Chute Deploy	10:14.3	10:14.8
Antenna Fairing Release	10:14.3	10:14.8
Main Chute Disconnect	14:47.7	15:22.0

Note: With the exception of Redstone booster cutoff, all events on the MR-3 flight were determined from commutated data. Thus the events could vary from the above times by as much as +0 to –0.8 seconds.

Appendix 4: Freedom 7 events and trajectory

Quantity	Planned	Actual
Range (N.M)	256.3	263.1
Maximum Altitude (N.M.)	100.3	101.2
Maximum Exit Dynamic Pressure lb/sq ft	598	586
Maximum Exit Longitudinal Load Factor, g	6.3	6.3
Maximum Reentry Dynamic Pressure lb/sq ft	591	605
Maximum Reentry Longitudinal Load Factor, g	10.8	11.0
Period of Weightlessness (Min:Sec)	04:53	05:04

Note: N.M. = nautical miles

About the author

Australian author Colin Burgess grew up in Sydney's southern suburbs. Initially working in the wages department of a major Sydney afternoon newspaper (where he first picked up his writing bug) and as a sales representative for a precious metals company, he subsequently joined Qantas Airways as a passenger handling agent in 1970 and two years later transferred to the airline's cabin crew. He would retire from Qantas as an onboard Customer Service Manager in 2002, after 32 years' service. During those flying years several of his books on the Australian prisoner-of-war experience and the first of his biographical books on space explorers such as Australian payload special Dr. Paul Scully-Power and teacher-in-space Christa McAuliffe had already been published. He has also written extensively on spaceflight subjects for astronomy and space-related magazines in Australia, the United Kingdom and the Unites States.

In 2003 the University of Nebraska Press appointed him series editor for the ongoing *Outward Odyssey* series of 12 books detailing the entire social history of space exploration, and he was involved in co-writing three of these volumes. His first Springer-Praxis book, *NASA's Scientist-Astronauts*, co-authored with British-based space historian David J. Shayler, was released in 2007. *Freedom 7* will be his sixth title with Springer-Praxis, for whom he is currently researching two further books for future publication. He regularly attends astronaut functions in the United States and is well known to many of the pioneering space explorers, allowing him to conduct personal interviews for these books.

Colin and his wife Patricia still live just south of Sydney. They have two grown sons, two grandsons and a granddaughter.

Index[1]

[1] Note: Index terms in italics are those that appear in a photo caption but not in the text.

A
ABMA. *See* Army Ballistic Missile Agency
Adams Elementary School, New Hampshire, 70
Adams Female Academy, New Hampshire, 70
Admiral Farragut Academy, New Jersey, 71
Advanced Research Projects Agency (ARPA), 5
6571st Aero Medical Research Laboratory, 33
Aerospace Medical Laboratory, Ohio, 41
The Airman (magazine), 31
American Heritage Magazine, 153
Amoskeag Savings Bank, 70
Andros Island, 149
Anne (sailing ship), 69
Antares Lunar Module, 239
Apollo missions, 237, 240
Apollo 14, 225, 239
Armel-Leftwich Visitor Center, 229
Armstrong, Neil, 86
Army Ballistic Missile Agency (ABMA), 1, 5, 7, 8
ARPA. *See* Advanced Research Projects Agency (ARPA)
Associated Press, 25, 26, 41, 109, 156, 185, 218–222, 241
Astronaut Scholarship Foundation, 240
Augerson, William, 47

B
Barka, David, 65, 68
Barka, Debi, 66, 68
Barka, Joe, 68
Barka, Nick, 68
Barka, Nicole, 68
Barka Oil Company, 66
Bartlett and Shepard Insurance Company, 70
Bartlett, Annie, 69
Beaver Falls, Pennsylvania, 179
Beddingfield, Sam, 121
Beneze, George, 72
Benson, Richard, 48, 55, 96
Bilderback, Jerry, 48
Blackshear, Hamilton, 31
Blunt, Barbara, 65
Brewer, Julia, 214
Brewer, Louise. *See* Shepard, Louise
Brewer, Phil, 214
Brewer, Russell, 185
Brucker, Wilber M., 5
Buckbee, Ed., 184
Burgess, Patricia, 259
Burke, Arleigh, 75
Burke, Walter F., 16
Byrnes, Martin, 162

C
California Science Center, 57
Cape Canaveral, Florida, 246
Cape Hatteras, North Carolina, 149
Carpenter, Scott, 8, 76–78, 82–84, 90, 100, 139, 233
Case Institute of Technology, 7
Chafee, John L.H., 241
Chesapeake Bay, Virginia, 165
Chimpanzees
  Caledonia, 33
  Chu, 33
  Duane, 33
  Elvis, 33
  Enos, 34
  George, 33

Index

Chimpanzees (*cont.*)
   Ham/Subject 65, 38
   Jim, 33
   Little Jim, 33
   Minnie/Subject 46, 41
   Paleface, 33
   Pattie, 33
   Roscoe, 33
   Tiger, 33
Chrysler Corporation, 3–5, 7
Churchley, Laura. *See* Shepard, Laura
Cohen, Arthur, 171
Communist Commercial Daily (newspaper), 214
Conger, Dean, 162, 167, 168, 176, 181, 183, 184, 186, 191, 192, 197, 199–201, 208, 209, 216, 217
Cooper, Gordon, 8, 77, 78, 81–84, 90, 92, 100, 117, 119, 131, 134, 136, 233
Cooper, Tom, 197
Cousins, Ralph W., 241
Cox, George, 48, 50, 164, 166, 167, 173, 176–178, 180, 181, 190, 201
Cox, Phillip, 211
Crispen, Paul, 56
Crowley, John W., 214

D
Dana, Bill, 90–92, 97, 129
Daniel, Allen, Jr., 23, 26, 166
Debus, Kurt, 19, 63, 141
Derry, New Hampshire, 65, 186, 241
Derry News (newspaper), 66
Derry Savings Bank, 70
Dittmer, Edward C., 29, 31–35, 41
Doner, Landis, 163, 198, 199, 202
Donlan, Charles, 75
D'Orsay, Leo, 196
Douglas, William ('Bill'), 79, 88, 103, 112–114, 116, 121, 129, 207, 210, 211
Dryden, Hugh L., 7, 74
Duncan, Russ, 197

E
Eisenhower, Dwight D., 7, 74
Eisman, Gene, 3
Emerson, Pauline Renza. *See* Shepard, Renza
Epic Rivalry (book), 3
Explorer 1 (satellite), 7, 117

F
Fédération Aéronautique Internationale, 170
First Parish Church, Derry, 68
Forest Hill Cemtery, 241
Fort Devens, Massachusetts, 69
Freeland, Lt., 162
French, Francis, 19, 57, 58, 150
Frost, Robert Lee, 65

G
Gagarin, Yuri, 99, 100, 103, 147, 170, 193, 213, 232
Gemini missions
   Gemini 3, xii
   Gemini 8, 86
General Dynamics Corporation, 80
General Electric Company, 1
George C. Marshall Space Flight Center, 7
Georgia University, 29
Gilruth, Robert R., 18, 23, 58, 59, 74, 93, 94, 221, 222
Glennan, T. Keith, 7, 74
Glenn, John, 8, 77, 78, 86, 90, 94, 95, 100, 103, 106, 108, 111–113, 122–125, 165, 208, 210, 227, 230, 233
Grand Bahama Island, 12, 23, 53, 150, 153, 154, 181, 185, 188, 193, 198, 207–212, 214, 249, 253, 254
Grant, Ulysses S., 69
Grass is Always Greener, The (movie), 214
Graveline, Duane, found in foreword xi, xiv
Great Lakes Training Center, Michigan, 189
Grenier Field, Manchester, 67, 70
Grissom, Virgil ('Gus'), 8, 23, 77, 78, 81, 82, 90, 94, 95, 100, 103, 108, 112, 114, 116, 117, 119, 121, 122, 124, 128, 165, 207–209, 213, 214, 222, 223, 233

H
Hackler, Ivan, 70
Hall, F.H.S., 162
Hall of Science and Exploration, 240
Hangar S (Cape Canaveral), 39, 53, 87, 89, 90, 100, 118, 196, 255, 256
Hardesty, Von, 3
Hawkins, Larry LaRue, 162
Hellriegel, John, 48, 52
Henry, James P., 33, 34, 41
Hickley, Don, 92
Hilles, Frederick V.H., 161

Holloman Aviation Medical Center, 31, 33
Hope, Bob, 91
House, Dr. William F., 238
HRH Duke of Edinburgh, 227
HRH Queen Elizabeth II, 227
Humphrey, John ('Jack'), 141
Huntsville Times (newspaper), 27

I
International Space Hall of Fame,
    New Mexico, 55, 57
'*Ivan Ivanovich*' (dummy cosmonaut), 63

J
Jackson, Carmault, 210, 211
Jenkins, Julie. *See* Shepard, Julie
Jet Propulsion Laboratory (JPL), 7
JFK Presidential Library and Museum, 235
Johnson, Lyndon B., 146, 216, 218
Johnson, Roy W., 5
Johnston, Richard S., 35
'José Jiménez' (Bill Dana character), 90–92, 117
JPL. *See* Jet Propulsion Laboratory (JPL)

K
Kansas Cosmosphere and Space Center, 229
Kapryan, Walter, 132
Keller, Kaufman T., 3
Kennedy, Jacqueline, 217
Kennedy, John F., 12, 58, 93, 103, 146, 191,
    214–218, 229, 233, 237
Khrushchev, Nikita, 213
Kiley, D.W., 162
Killian, Ed., 159–161, 163, 164, 168–171, 179,
    182, 186, 188, 190, 194–199, 201–204
King, John ('Jack'), 109
Koch, George P., 161, 193
Koons, Wayne, 23, 26, 164–167, 173, 175, 176,
    179, 180, 190, 201–203
Korabl-Sputnik 5 (satellite), 63
Kraft, Christopher C., 47, 97, 131, 132, 141, 144, 214
Kreitzberg, Larry, 178, 190, 215
Küttner, Joachim, 11

L
Laika (dog), xxv
Lake Okeechobee, Florida, 149

Langley Research Center, Virginia, 59,
    74, 227
Laning, Robert C., 180, 184
Levy, Edwin, 75
Lewis Research Center, Ohio, 83
Life (magazine), 144, 176
Light This Candle (book), 135
Lindbergh, Charles, 70
Lindell, Keith, 80
Little Jim, 33
Louisiana Polytechnic Institute, 3
Lovelace Clinic, New Mexico, 74, 77
Lovell, James, 77
Lowe, Nancy, 77
Low, George M., 16
Luetjen, H.H. ('Luge'), 12

M
MAC. *See* McDonnell Aircraft Corporation
    (MAC)
Marine Air Group 252, 23
Marshall, George C., 3, 7, 27
Maxwell, Rev. Henry, 131
Mayflower (sailing ship), 69
May, Hugh, 207
McAuliffe, Christa, 259
McCool, Alex, 27
McCormack, John, 221
McDonnell Aircraft Corporation (MAC), 7, 13,
    18–20, 30, 59, 78, 255
McGivern, Anissa, 68
McGivern, Finnegan, 68
McGivern, Mike, 68
McKay, Jean, 113
McLane, A.W., 162
Mercury 7 Foundation, 240
Mercury-Redstone missions, 94
    MR-1, 12–20
    MR-2, 29, 30, 35, 36, 38–41, 45, 47, 48, 52,
        57–59, 64, 96, 165, 166, 168
    MR-3, 35, 59, 97, 103, 108, 109, 127, 133,
        139, 142, 163, 165, 166, 169, 236,
        251–254
    MR-1A, 20–27, 166
    MR-BD, 59–63
Mercury spacecraft
    Freedom 7, 59, 68, 96, 98, 106, 124, 125
    Freedom 7 II, 237, 238
    Liberty Bell 7, 213, 222
Meredith, James, 29

Index

Miss Sam (monkey), 60, 61
Mitchell, Edgar, 239
Mittauer, Richard, 163, 194
Molnoski, Paul, 189
Monterey Community Hospital, California, 241
Moon Shot (book), 139, 218
Mosely, Dan, 38, 39, 41
Mundt, Karl, 193
Munger, Bob, 90

N
NAA. *See* North American Aviation (NAA)
NACA. *See* National Advisory Committee for Aeronautics (NACA)
NASA. *See* National Aeronautics and Space Administration (NASA)
National Academy of Sciences, 245
National Advisory Committee for Aeronautics (NACA), 7
National Aeronautics and Space Act, 74
National Aeronautics and Space Administration (NASA), 5–11, 16, 23, 25, 28, 33, 39, 41, 47, 59–61, 65–102, 105, 106, 108–110, 121, 141, 154, 162–166, 168–170, 173, 175, 180, 183–187, 192–194, 204, 212, 213, 215–217, 219, 221, 222, 225, 229, 237–240, 244, 247, 250–253
National Air and Space Museum, 227, 228, 237
National Geographic (magazine), 162
National Institutes of Health, 245
National Zoo, Washington, D.C., 55
Naval Aviation News (magazine), 65
Naval War College, Rhode Island, 73
Neal, Roy, 93
Nixon, Richard M., 12, 189
Norman, R.J., 39, 162
North American Aviation (NAA), 3, 7
North Carolina Zoological Park, 55
North, Warren, 75
Nottingham, New Hampshire, 69

O
Oak Street School, Derry, 70
OGMC. *See* Ordnance Guided Missile Center (OGMC)
O'Hara, Delores ('Dee'), 88–90, 114, 116, 207, 210
Operation Hardtack, 5
Ordnance Guided Missile Center (OGMC), 1, 3
Ottawa University, 165
Outward Odyssey (book series), 259

P
Patillo, W.H., 162
Pebble Beach, California, 241
Pesman, Gerard J., 35
Pickett, Andy, 136
Pinkerton Academy, New Hampshire, 70–72
Poulis, Victor, 238
Powell, Welsey, 67
Powers, John ('Shorty'), 110, 170, 210, 216
Project Adam, 5, 7
Project Apollo, xxvii, 236
Project Man Very High, 5
Project Mercury, 1, 5, 7, 8, 16, 18, 34, 75, 77, 100, 162, 210, 237, 245, 246, 250
Project Mercury Control Center, 246

Q
Qantas Airways, 259

R
Rayburn, Sam, 221
Redstone Arsenal, Huntsville, 1, 3, 5, 7, 80
Remar, Jim, 230
Rice University, Houston, 229
Richmond, Michael, 189, 190
Right Stuff, The (book), 77, 93, 136
Roberts, O.A., 162
Rocketdyne AQ-7 (engine), 7, 12
Rocketdyne Division, NAA, 3, 7
Rogers, Buck, 70
Roosa, Stuart, 239
Roosevelt, Franklin D., 69
Royal Scottish Museum, Edinburgh, 227, 234
Ruff, George, 75, 211

S
Schirra, Wally, 8, 44, 77, 78, 83, 90, 92, 100, 139, 150, 210, 233
Schmitt, Joe, 105, 106, 114, 116, 117, 124
Science Museum, London, 227, 231–233
Scott, Dave, 86
Scully-Power, Paul, 259
Shayler, David J., 259
Shepard, Alan B. (Sr.), 69, 70, 79
Shepard, Alan B., Jr., 65, 68, 69, 233, 242, 245–250
Shepard, Alice, 185
Shepard, Frederick J., 68
Shepard, Frederick Jr., 68
Shepard, Henry, 68
Shepard, Julie, 72, 241

Shepard, Laura, 72, 241
Shepard, Louise, 72, 74, 94, 114, 134, 185, 208, 214, 216, 218, 219, 241, 242
Shepard, Pauline ('Polly'), 68, 69, 186, 216
Shepard, Renza, 69, 73, 79, 130, 214, 215
Sherman, David, 65
Sherman, Polly. *See* Shepard, Pauline ('Polly')
Sidonia (ship), 227
Simons, David, 33
Skidmore, Howard, 159, 160, 162, 163, 169, 183, 190
Slayton, Donald ('Deke'), 8, 23, 44, 77–80, 90, 93, 100, 129, 131, 136, 139, 141, 150, 153, 207, 208, 245, 248, 251–254
Smithsonian Institution, Washington, D.C., 227–229
Space Shuttle *Enterprise*, 189
Space Task Group (STG), 7, 8, 23, 33, 59, 74, 75, 78–80, 93, 165, 168, 246, 250
Spirit of St. Louis (airplane), 240
Sputnik (satellite), 63, 100
Stafford, Tom, 237, 238
STG. *See* Space Task Group (STG)
Stringely, Norman, 39
Strong, Jerome, 180, 183, 184
Szathmary, William. *See* Dana, Bill

T
Taylor, Robert H., 162
Thiokol Chemical Company, 3
This New Ocean (book), 16, 162
Thompson, Neal, 241
Thompson, Robert, 162
Thompson, Scott, 179
Thornton, Matthew, 65
Time (magazine), 213
Titov, Gherman, 225, 227, 230
Tynan, Charles, 162, 163, 175, 180, 187, 188, 194–196, 198, 199, 202

U
Udvar-Hazy Center, Washington, D.C., 237
United States Ship (USS)
 Abbot (DD-629), 162
 Ability (MSO-519), 162
 Borie (DD-215), 48
 Cogswell (DD-651), 72
 Decatur (DD-936), 162
 Donner (LSD-20), 48, 51
 Ellison (DD-454), 48, 49
 Lake Champlain (CVS-39), 131, 159–162, 167, 169, 178, 180, 189, 193, 198, 203, 204, 211, 215, 249
 Manley (DD-940), 48
 Newman K. Perry (DD-883), 162
 Notable (MSO-460), 162
 Recovery (ARS-43), 162
 Rooks (DD-804), 162
 The Sullivans (DD-537), 162
 Valley Forge (CV-45), 23, 25, 166
 Wadleigh (DD-689), 162
University of Mississippi, 29
University of Nebraska Press, 259
U.S. Air Force Bases
 Edwards AFB, California, 80
 Holloman AFB, New Mexico, xi, 33, 34, 54
 Langley Field, Virginia, 23, 34, 77, 86, 246
 Patrick AFB, Florida, 53, 88, 196
 Stead AFB, Nevada, 87
 Vandenberg AFB, California, 80
U.S. Marine Corps Air Stations, 164, 165
 MCAS New River, North Carolina, 164
U.S. Naval Academy, Annapolis, 71, 227, 229, 235
U.S. Naval Air Bases
 NAS Mayport, Florida, 159, 160, 162
 NAS Patuxent River, Maryland, 72, 73
 NAS Quonset Point, Rhode Island, 159, 160, 162
U.S. Naval Air Development Center, 80, 245
U.S. Navy Test Pilot School, Maryland, 72
U.S. Rockets
 Atlas, xxvi, xxvii, 80, 90
 Corporal, 3, 168
 Hermes A3, 3
 Hermes C1, 3
 Jupiter-C, 7, 8
 Little Joe, 59, 60
 Redstone, 1, 7, 12, 14, 16, 20, 24, 26, 29, 43, 96, 99, 141, 143, 154
USS. *See* United States Ship (USS)
U.S. Space and Rocket Center, Alabama, 184

V
Vitulli, Anthony, 189
Voas, Robert, 75, 86, 211
von Braun, Wehrner, 1, 3, 7, 8, 11, 23, 27, 58, 59, 80, 96, 134, 140, 141

W
Wackermann, Alice. *See* Shepard, Alice
Walker, Elizabeth, 68
Wallops Island, Virginia, 12, 60, 61
Warren, Ann, 68
Warren, Joseph, 68

Warren, Mary, 68
Warren, Nathaniel, 68
Warren, Richard, 68
Warren, Sarah, 68
Webb, James, 58, 59, 221, 227, 237
Wendt, Guenter, 18–20, 26, 90, 100
We Seven (book), 71, 148, 154
Weymouth, Ralph, 159, 160, 169, 170, 193–195, 198, 201, 202
Wheelwright, Charles, 39
White, Stanley C., 34, 35
White, Thomas, 75
Wiesner Committee, 58
Wiesner, Jerome B., 58
Wiggins, Bertha, 70
Wilbur, Ted, 198

Wiley, Alexander, 193
Williams, Allen, 3, 4
Williams, Walter, 63, 131, 132
Wolfe, Tom, 74, 81
Wright Air Development Center, Ohio, 74, 81
Wright brothers, 69

Y
Yaquiant, Frank, 204
Yardley, John, 19

Z
Zeiler, Albert, 141
Zvezdochka (dog), 63

CPSIA information can be obtained at www.ICGtesting.com
Printed in the USA
LVOW11s1630021013

355120LV00005B/101/P

9 783319 011554